THE PRINCIPLES
OF
THERMODYNAMICS

THE PRINCIPLES
OF
THERMODYNAMICS

by

GEORGE BIRTWISTLE,

FELLOW OF PEMBROKE COLLEGE,
CAMBRIDGE

CAMBRIDGE
AT THE UNIVERSITY PRESS
1927

CAMBRIDGE UNIVERSITY PRESS
Cambridge, New York, Melbourne, Madrid, Cape Town,
Singapore, São Paulo, Delhi, Mexico City

Cambridge University Press
The Edinburgh Building, Cambridge CB2 8RU, UK

Published in the United States of America by Cambridge University Press, New York

www.cambridge.org
Information on this title: www.cambridge.org/9781107660434

© Cambridge University Press 1927

First edition 1925
Second edition 1927
First published 1927
First paperback edition 2013

A catalogue record for this publication is available from the British Library

ISBN 978-1-107-66043-4 Paperback

PREFACE

THIS book contains the substance of lectures given in the University of Cambridge in the summer terms of the past two years to men whose future interest may have been any of mathematics, physics, chemistry, astronomy or mechanical science. The object was to set out with care the foundation principles of the subject and to illustrate them by applications to these various branches of science, where no more than a general knowledge of the science was required for their appreciation. For this reason the relation of statistical mechanics to thermodynamics, and questions, such as the connection between magnetism and temperature, requiring a special knowledge of dynamical or physical theory, were not included.

G. B.

EIDSLAND
  BERGEN
    NORWAY

1 *October* 1924

PREFACE TO THE SECOND EDITION

This edition has been revised and a chapter on Nernst's heat theorem added.

G. B.

1 *October* 1927

CONTENTS

PAGES

CHAPTER I. PRELIMINARIES 1–9

The dynamical theory of heat—Measurement of temperature—The laws of Boyle, Charles, and Avogadro—Perfect gas—Indicator diagram.

CHAPTER II. THE TWO LAWS OF THERMODYNAMICS; THE CARNOT CYCLE 10–24

The first law—Joule's experiment—Specific heat—Properties of a perfect gas—Isothermals and adiabatics—Carnot's cycle—Reversibility of the cycle—The second law—Carnot's principle—Efficiency of a reversible engine—Thomson's scale of temperature—Carnot's cycle for a perfect gas.

CHAPTER III. DISSIPATION OF MECHANICAL ENERGY; ENTROPY 25–38

The Carnot-Clausius equation—The dissipation of mechanical energy in natural processes—Entropy—Increase of entropy in natural processes—Irreversible engine—Entropy and probability—Entropy-temperature diagram—Steam tables.

CHAPTER IV. THERMODYNAMICS OF A FLUID; CHANGE OF STATE 39–51

Vapours—Isothermals of a liquid and its vapour—Continuity of the liquid and gaseous states—Liquefaction of gases—Ideal isothermal of a fluid—Van der Waals' equation—The reduced form of the equation—Boiling point and pressure—Amagat's experiments—Various forms of characteristic equation due to Clausius, Dieterici, Lees, Callendar, Berthelot.

CHAPTER V. THERMODYNAMIC FUNCTIONS . . 52–58

Internal energy—Total heat—The thermodynamic potentials ψ, ζ—The Thomson equation—Boundary curves for steam and water.

CHAPTER VI. THE RANKINE CYCLE; REFRIGERATION 59–67

The Rankine cycle—Efficiency of the cycle—Calculations for dry, wet and superheated steam—The use of the function G—Mollier diagram—Refrigeration.

PAGES

CHAPTER VII. THERMODYNAMICAL RELATIONS
FOR A SIMPLE SUBSTANCE   68–75
Theorems on partial differentiation—Maxwell's equations—Specific heat equations—Variation of c_v with
volume—Further properties of specific heats—Calculation of E, ϕ, etc. for a simple system.

CHAPTER VIII. THE JOULE-THOMSON POROUS
PLUG EXPERIMENT   76–85
Theory of the experiment—Calculation of the cooling
effect by the use of Van der Waals' equation—Inversion
of the effect—Deduction of the characteristic equation
of a gas from observations of the cooling effect—Reduction of the readings of an air thermometer to
Thomson's scale and the determination of the thermodynamic zero of temperature—Linde's process for the
liquefaction of air.

CHAPTER IX. CLAPEYRON'S EQUATION; CLAUSIUS'
EQUATION.   86–94
Vapour pressure and temperature—Clapeyron's equation—Deduction from the ζ-function and by the use of
a Carnot cycle—Effect of pressure on the boiling point
and the melting point—Numerical results—Clausius'
equation—Specific heat of saturated steam.

CHAPTER X. EQUILIBRIUM OF SYSTEMS; THE
PHASE RULE   95–107
Conditions of natural change of a system—Isothermal
changes—Conditions of equilibrium—Fundamental
equation of a simple substance—Thermodynamic surface for a simple substance—Geometrical condition of
stability—Homogeneous mixtures—Fundamental equations for such a mixture—Heterogeneous mixtures—
Gibbs' conditions of equilibrium at constant temperature and pressure—Proof of the phase rule—Illustrations of the phase rule—Thermodynamic surface for a
homogeneous mixture of two components—Distillation
of a mixture of two liquids—Alloys.

CHAPTER XI. OSMOTIC PRESSURE; VAPOUR PRES-
SURE; GAS MIXTURES   108–122
Osmotic pressure of solutions—Osmotic pressure of
gases—Vapour pressure and osmotic pressure—Vapour
pressure and concentration of a solution—Boiling point
of a solution—Determination of molecular weight of a
dissolved substance from the rise of the boiling point—
Freezing point of solutions—Gas mixtures and dilute
solutions—Entropy of a gas mixture—Planck's results
—Surface tension and temperature.

PAGES

CHAPTER XII. THERMOELECTRIC PHENOMENA .   123–128
The Seebeck, Peltier, and Thomson effects—Theory of
the thermoelectric circuit—Reversible cell—The E.M.F.
of a Daniell cell—Theory of a reversible cell.

CHAPTER XIII. GAS THEORY; VARIATION OF
SPECIFIC HEAT WITH TEMPERATURE   .   .   129–145
Elements of gas theory—Atoms and molecules—
Rutherford's atomic model—The rise of specific heat
of a gas with temperature—Use of quantum mechanics
—Application to gas-engine theory—Specific heat of
solids at low temperatures; Debye's formula—Atomic
heat near the absolute zero.

CHAPTER XIV. RADIATION   146–163
Nature of radiation—Pressure of radiation—Stefan's
law and Boltzmann's theoretical deduction of the law—
Deduction by Thomson's equation—Experimental verification of Stefan's law—The spectrum of a black body—
The effect of change of temperature on the distribution
of energy in the spectrum of a black body—Wien's displacement law and the formula $\lambda^{-5} F (\lambda t)$—Experimental
verification of Wien's law—Distribution of energy in the
spectrum; Planck's formula—Stellar spectra—Stellar
evolution—Eddington's theory of the interior of a star.

CHAPTER XV. THE THIRD LAW OF THERMO-
DYNAMICS   164–168
Kirchhoff's equation—Nernst's heat theorem.

THE PRINCIPLES OF
THERMODYNAMICS

CHAPTER I

PRELIMINARIES

1. *The dynamical theory of heat.* Down to the end of
the eighteenth century, heat was generally regarded as
substance of no appreciable mass which could pass into or
out of the space between the ultimate particles of a body.
This substance, 'caloric' as it was called, could neither be
created nor destroyed. At that time, too, there were many
who regarded heat as an effect due to the motion of the
particles themselves.

The experiments of Davy (1799) on the melting of two
pieces of ice by friction between them, and the observations
of Rumford[1] on the great heat of the shavings produced in
the boring of cannon were in themselves sufficient to show
that heat could be generated by mechanical effort, but they
were not then fully appreciated.

In the decade 1840–50 the work of three or four experi-
menters, of whom Joule was pre-eminent, showed that the
heat developed from mechanical work was in proportion
to the work spent. In 1843[2] Joule showed that in order to
produce a unit of heat (the heat required to raise the tem-

---

[1] *Phil. Trans.* 1798.
[2] *Phil. Mag.* 1843.

perature of a pound of water 1° F.) the mechanical work
expended was 770 foot-pounds.

The number of units of work expended in producing a
unit of heat is called the 'mechanical equivalent of heat.'

More recent determinations give the following results:

> 1400 ft.-lb. of work are equivalent to 1 lb.-calorie
> (the heat required to raise the temperature of a pound
> of water from 0° C. to 1° C.),

or 41.8×10^6 ergs of work are equivalent to 1 gram-
calorie (the heat required to raise the temperature of
a gram of water from 0° C. to 1° C.).

In 1847 Helmholtz's tract *Über die Erhaltung der Kraft*
appeared, in which he asserted the principle of the conservation of energy and developed the ramifications of that
principle throughout natural phenomena. Of the forms of
energy, Joule's experiments enabled him to count heat as
one; the reception of heat by a body meant an increase in
the energy of its ultimate particles.

Though Helmholtz accepted Joule's principle, Thomson
(who in 1847 was the first to see the importance of Joule's
work) was at this time unable to accept or reject it, because
of its apparent conflict with Carnot's principle.

In 1824 Sadi Carnot published an essay on *La puissance
motrice du feu*, but it was forgotten until 1845 when
Thomson, who had learnt of its existence through an
account of it in a memoir by Clapeyron[1] in 1834, realised
its profound importance—'an epoch-making gift to science,'
he afterwards declared. In this essay Carnot endeavours
to find how it is that heat produces mechanical effect in
an engine. He considers a 'cycle' of operations, at the end
of which the working substance of the engine is left in the
same physical condition as it was at the beginning, this
being essential if the mechanical effect is to be assigned
solely to the heat given to the substance. He assumes that

[1] *Jour. École Polyt.* xiv.

since the state of the substance is unaltered by the cycle, the heat received by it from the boiler is equal to the heat given out by it to the condenser, and concludes that the heat produces mechanical work by falling from one level of temperature to another, just as water in falling from one level to another over a water-wheel can do work in proportion to the quantity which descends and the height through which it falls.

He proceeds to describe an ideal cycle of operations (now known as Carnot's cycle) which has the property of being 'reversible,' and to prove that no engine receiving heat at one temperature and emitting heat at a lower temperature can produce more mechanical work from a given quantity of heat than an engine describing a reversible cycle between the same two temperatures. The proof depends upon the equality of the heat received and rejected by the engine.

In 1849, James Thomson[1], using Carnot's ideas, proved theoretically that the freezing point of water must be lowered by pressure, and calculated that the lowering of the freezing point in degrees Centigrade due to a pressure of n atmospheres in excess of atmospheric pressure should be $(\cdot 0075)\, n$.

This result was confirmed by W. Thomson's experiments in 1850[2] and strengthened his belief in Carnot's principle.

It was now (1850) that Clausius[3] resolved the difficulty of reconciling Carnot's principle (which assumed production of mechanical work without loss of heat) with the views of Joule and Helmholtz (which required equivalence between mechanical work and heat). His view was that in an engine less heat is given out at the lower level of temperature than is taken in at the upper level of temperature, and that the difference is converted into mechanical work. This con-

---

[1] *Trans. R.S.E.* Jan. 1849.
[2] *Proc. R.S.E.* Jan. 1850.
[3] *Pogg. Ann.* 1850.

clusion had been reached independently by Thomson[1] and was published by him in 1851.

There was still the difficulty that Carnot's proof of reversibility as the test of the most effective engine depended on the equality of the heat received at the higher temperature and the heat rejected at the lower. Clausius showed that if Carnot's principle that a reversible engine is the most efficient was to be retained with the theory he proposed, a new axiom was required, which is now known as the second law of thermodynamics. This axiom, and the principle of conservation of energy, form the foundation of modern thermodynamics.

2. *Measurement of temperature.* In a perfect thermometer equal elevations of temperature, as indicated by the divisions of the scale, correspond to equal increases in volume of the substance. As the coefficients of expansion of the substance and the material enclosing it vary with the temperature, equal elevations of temperature do not correspond to equal increases in volume of the substance.

Thus the readings of an actual thermometer depend upon the physical properties of the materials, such as mercury, air, glass, used in its construction.

It is found that if one of the 'permanent' gases is used the expansion is so uniform over a large range of temperature that the indications of such a thermometer, for example an air thermometer, are for most practical work perfect enough, though they would not be so for very low temperatures, in the neighbourhood of the point of liquefaction of the gas where the variation of expansion becomes apparent.

Lord Kelvin (Sir W. Thomson) in 1848[2] perceived the need of a scale of temperature independent of the properties of any particular substance and that in Carnot's principle he had the means to establish such a scale, the

---

[1] *Trans. R.S.E.* March 1851.
[2] *Proc. Camb. Phil. Soc.* June 1848.

absolute thermodynamical scale of temperature, which in its final form was given in his memoir on the 'Dynamical Theory of Heat[1].'

3. *The laws of Boyle and Charles.* These are experimental laws which hold approximately for gases and vapours and become more and more exact the further the gas or vapour is from its point of liquefaction.

Boyle's law is 'The pressure of a given mass of a gas at constant temperature varies inversely as the volume.'

Gases expand at constant pressure according to the law $v = v_0 \left(1 + \dfrac{\theta}{273}\right)$ where v is the volume at $\theta°$ C. and v_0 is the volume at $0°$ C., so that if v_1; v_2 are the volumes at temperatures $\theta_1°$ C., $\theta_2°$ C., $\dfrac{v_2}{v_1} = \dfrac{273 + \theta_2}{273 + \theta_1} = \dfrac{t_2}{t_1}$, where $t = 273 + \theta$. Temperatures t, obtained by adding 273 to the Centigrade temperature θ, are called 'absolute temperatures on the gas thermometer scale.' We thus have Charles' law, which is 'The temperature of a given mass of gas at constant pressure varies directly as the volume.'

These laws are included in the formula $pv = at$, where v is volume of 1 gram of the gas (its specific volume) and a is a constant for a given gas.

4. *Avogadro's law* is 'All gases at the same temperature and pressure contain the same number of molecules per unit volume.' Modern work on the electrical properties of gases shows that this number (*Avogadro's number*) is $2 \cdot 70 \times 10^{19}$ per cubic centimetre[2] at $0°$ C. and atmospheric pressure (760 mm. of mercury).

5. *The gas constant.* Now let n be the number of molecules per c.c. and m be the mass of a molecule of a given

---

[1] *Trans. Roy. Soc. Edin.* xxi. 1. 1854, or *Math. and Phys. Papers*, I. p. 235.

[2] A. Sommerfeld, *Atomic Structure and Spectral Lines*, p. 535.

gas in grams. Then mn is the mass in grams of 1 c.c. and $\dfrac{1}{mn}$ is the volume in c.c. of 1 gram.

Hence $pv = at$ becomes $p \cdot \dfrac{1}{mn} = at$,

or $\qquad p = n\,(ma)\,t = nRt$, where $R = ma$.

Since, by Avogadro's law, n is the same for all gases at the same temperature and pressure, R is the same constant for all gases.

Using Avogadro's number given above, R can be calculated. For, a pressure of 760 mm. of mercury

$$= 981 \times 13\cdot6 \times 76 \text{ dynes per sq. cm.}$$
$$= 1013600 \text{ dynes per sq. cm.}$$
$$\therefore\ R = \frac{1013600}{2\cdot7 \times 10^{19} \times 273} = 13\cdot8 \times 10^{-17}.$$

This is the 'universal gas constant' used in the kinetic theory of gases.

In physical chemistry, a unit of mass called a 'gram-molecule' or a 'mol' is very generally used. 1 mol of a substance is M grams, where M is the 'molecular weight' of the substance in the chemical sense. Thus the molecular weight of hydrogen being 2 (1·008) or 2·016, 1 mol of hydrogen is 2·016 grams; the molecular weight of oxygen being 2 (16) or 32, 1 mol of oxygen is 32 grams.

Now let n' be the number of mols per c.c.

Then $n'M$ is the number of grams per c.c. and this is equal to nm,

$$\therefore\ \frac{n}{n'} = \frac{M}{m}.$$

Now the molecular weight is proportional to the mass of a molecule, so that $\dfrac{M}{m}$ is the same for all gases, and therefore $\dfrac{n}{n'}$ is the same for all gases, and equal to N, say.

This ratio N is the number of molecules per mol and its value can be found by using the constants for oxygen: 1 c.c. of oxygen at $0°$ and 760 mm. weighs ·001429 gram and 1 mol of oxygen is 32 grams,

$$\therefore n' = \frac{·001429}{32} = 44·65 \times 10^{-6}.$$

Now $n = 2·7 \times 10^{19}$,

$$\therefore N = \frac{n}{n'} = \frac{2·7 \times 10^{19}}{44·65 \times 10^{-6}} = 6·06 \times 10^{23}$$

(*Avogadro's number per mol*[1]).

Now $p = nRt = n'NRt = n'R't$,

where $\qquad R' = NR,$
$$= (6·06 \times 10^{23})(13·8 \times 10^{-17}),$$
$$= 83·6 \times 10^{6}.$$

R' is also the same for all gases, and is the 'universal gas constant' used in physical chemistry.

Further, if v' is the volume of 1 mol of the gas, the number of mols per c.c. is $\frac{1}{v'}$,

$$\therefore \frac{1}{v'} = n', \text{ and } p = n'R't \text{ becomes } pv' = R't.$$

Again $pv = at$, where v = volume of 1 gram of the gas,

$$\therefore v' = Mv,$$
$$\therefore pv' = Mat.$$

Hence $\qquad R' = Ma$ or $a = \dfrac{R'}{M} = \dfrac{83·6 \times 10^{6}}{M};$

$$\therefore pv = at, \text{ where } a = \frac{83·6 \times 10^{6}}{M}.$$

These results may be summarised as follows:

I. $p = nRt$,

where (i) if n is the number of molecules per c.c., $R = 13·8 \times 10^{-17}$, and (ii) if n is the number of mols per c.c., $R = 83·6 \times 10^{6}$, for all gases.

[1] Cf. du Noüy, *Phil. Mag.* Oct. 1924.

II. $pv = Rt,$

where (i) if v is the volume of 1 mol of gas, $R = 83.6 \times 10^6$,

and (ii) if v is the volume of 1 gram of gas, $R = \dfrac{83.6 \times 10^6}{M}$,

where M is the molecular weight of the gas.

In the above, pv has the dimensions of work, so that R is given in ergs.

Now 41.8×10^6 ergs of work produce 1 gr.-calorie of heat, so that if v is the volume of 1 mol of gas,

$$pv = (83.6 \times 10^6) \cdot t \text{ ergs},$$

$$= \frac{83.6 \times 10^6}{41.8 \times 10^6} \cdot t \text{ gr.-calories},$$

$$= 2t \text{ gr.-calories}.$$

Thus using heat units, $pv = 2t$, where v is the volume of 1 mol.

6. *Perfect gas.* This is an ideal gas, infinitely far from the point of liquefaction, of which one of the properties is that of satisfying Boyle's and Charles' laws exactly, so that $pv = Rt$.

7. *Characteristic equation.* The equation $pv = Rt$ for a perfect gas is an example of the equation $f(p, v, t) = 0$ which holds for any 'simple' substance, such as a homogeneous fluid. Such an equation is called the 'characteristic' equation of the substance. If any two of the three variables p, v, t which determine the physical state of the substance are known, the third can be found from the characteristic equation.

Hence the 'state' of the substance is determined by any two of p, v, t.

8. *Indicator diagram.* If p, v are taken as ordinate and abscissa of a point A, this point indicates the state of the substance. If the substance passes to another state B, the changes it undergoes during the passage are shown by a

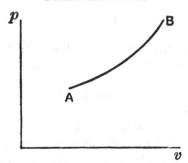

curve joining A to B. The figure is an 'indicator diagram' of the kind first used by Watt to indicate the state of the steam in the cylinder of an engine.

9. *Work done by a fluid in expansion.* Let S and S' represent the surface of a fluid before and after a small expansion against an external pressure p which is constant over the surface, the pressure of the substance being supposed infinitesimally greater than the external pressure, which makes the expansion possible and a slow one so that no energy of motion is developed.

Consider the element dS of the surface and let its displacement along the normal be dn. The work done by the substance is then $\Sigma\,(pdS)\,dn$

$$= p\Sigma dS \,.\, dn = p \text{ (increase of volume)} = p\,dv.$$

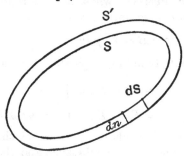

CHAPTER II

THE TWO LAWS OF THERMODYNAMICS;
THE CARNOT CYCLE

10. *The first law of thermodynamics.* This is the principle of the conservation of energy expressed so as to include the energy of heat. A body in a given state has 'internal energy' associated with the configuration and motion of its molecules. The change of its internal energy when it passes from one state to another is independent of the path between the two states and depends only on the configuration and motion in the respective states. The difference between the internal energy in any state and that in some standard state will be denoted by E.

If the substance takes in a quantity of heat Q (measured in units of work) and performs W units of external work, let the internal energy change from E_1 to E_2. Then by the conservation of energy,

$$E_2 - E_1 = Q - W.$$

This is the first law of thermodynamics—in words, it is 'The heat taken in by a substance

= the increase of its internal energy

+ the work done *by* the substance.'

For a simple substance whose state is represented by a point on the p-v diagram, the work done in passing from a state A to a state $B = \int_{v_A}^{v_B} p\, dv$.

Hence the heat taken in, in passing from A to B,

$$= E_B - E_A + \int_{v_A}^{v_B} p\, dv,$$

$$= E_B - E_A + (\text{area } ABML).$$

Since this area depends on the form of the curve AB, both the work done by the substance and the heat taken in depend on the path from A to B.

Thus the quantity of heat which is taken in by a body in passing to any state from some standard state cannot be expressed as a function of the state; it may have any value depending on how that state was reached from the standard one.

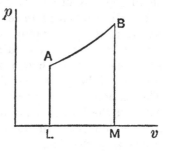

If dE is the small increase of internal energy which accompanies the absorption of a small amount of heat dQ and the performance by the substance of a small amount of work dW, then

$$dE = dQ - dW.$$

But it is understood that whilst dE is the differential of a function E of the variables which determine the state, dQ and dW are not differentials of functions Q, W of the state, as such functions do not exist.

11. *Joule's experiment on the expansion of a gas into a vacuum*[1]. Joule compressed air to about 20 atmospheres in a strong vessel which was connected by a pipe, containing a stop-cock, to another vessel previously exhausted. The whole was placed in a vessel of water.

On opening the stop-cock, the air rushed from the first vessel to the second so that in a short time the pressure was the same in both. On measuring the temperature of the water again, no change was perceptible.

In this experiment no external work was done by the air and no heat entered or left. Hence by the first law, the internal energy of the air was unchanged by the sudden expansion. Now the internal energy E is a function of the state and is in general a function of two of p, v, t, so that E is a function of v and t. But E is unaffected by change of volume and is therefore a function of t only.

[1] *Phil. Mag.* 1845.

This experiment was not a very delicate one, on account of the large volume of water used, which might absorb a small quantity of heat without a sensible rise of temperature. By a much more sensitive variant of this method, known as the porous plug experiment, Joule and Thomson during 1852–62 carried out a long series of determinations in which a small change of temperature was shown to occur in all the gases used (air, carbonic acid and hydrogen). The change was much less in the case of hydrogen than of air, and the former being much further from its point of liquefaction than air at ordinary temperatures is more nearly a 'perfect' gas.

It will be assumed, then, that a 'perfect' gas has also the property that in the Joule experiment, its change of temperature would be zero exactly. Hence the internal energy of a perfect gas is a function of its temperature only.

12. *Specific heat.* Let the temperature of 1 gram of a substance rise from t to $t + dt$ owing to the absorption of a small quantity of heat dQ and let $\dfrac{dQ}{dt} = c$. Then when dt is infinitesimal, c is called the 'specific heat' of the substance at temperature t.

Any number of specific heats may be defined at a given temperature, according to the conditions under which the heating takes place. For it has been seen that the quantity of heat taken in by a substance in a given change of state depends on the path by which that change is effected.

Thus if the pressure is constant during the change $\dfrac{dQ}{dt}$ has the value c_p, which is 'the specific heat at constant pressure,' and if the volume is constant during the change $\dfrac{dQ}{dt}$ has the value c_v, 'the specific heat at constant volume.'

Now
$$dQ = dE + p\,dv,$$
$$\therefore\ c = \frac{dE}{dt} + p\,\frac{dv}{dt}.$$

If v is constant, $c = c_v$,

$$\therefore \quad c_v = \frac{dE}{dt}.$$

Now for a perfect gas E is a function of t only; therefore c_v is a function of t only.

For the permanent gases, c_v is independent of t except for very high or very low temperatures, so that a further condition will be assumed for a perfect gas, viz. that c_v is constant for all temperatures.

Hence $\dfrac{dE}{dt} = c_v$ and $E = c_v t$ for a perfect gas, if E is taken to be 0 when $t = 0$.

13. *Properties of a perfect gas.* The gas satisfies the conditions $pv = Rt$, $E = c_v t$.

Now
$$dQ = dE + p\,dv,$$
$$= c_v dt + p\,dv,$$

and
$$\frac{dp}{p} + \frac{dv}{v} = \frac{dt}{t},$$

$$\therefore \quad dQ = c_v dt + pv\left(\frac{dt}{t} - \frac{dp}{p}\right),$$
$$= c_v dt + R\,dt - v\,dp.$$

But $dQ = c_p dt$ when $dp = 0$,

$$\therefore \quad c_v + R = c_p,$$

or
$$c_p - c_v = R.$$

Hence c_p is also constant for a perfect gas, and $c_p > c_v$.

Also
$$dQ = c_v dt + p\,dv,$$
$$= c_p dt - v\,dp.$$

14. *Adiabatic expansion.* This is an expansion of a substance, which may be considered as contained in a vessel fitted with a movable piston, when the walls of the vessel and the piston are impermeable to heat.

The expansion is supposed to be carried out against an external pressure sufficiently slowly for the substance to

be considered in equilibrium with the external pressure at
any stage of the process (the process in fact is to be one
which later will be described as 'reversible').

Under these conditions,

$$dQ = 0,$$

and in the case of a perfect gas

$$\left.\begin{array}{l} c_v dt + p\,dv = 0 \\ c_p dt - v\,dp = 0 \end{array}\right\},$$

$$\therefore \frac{c_p}{c_v} = -\frac{v\,dp}{p\,dv}.$$

If $\dfrac{c_p}{c_v}$, which is constant, is denoted by γ, $(\gamma > 1)$,

$$\gamma \frac{dv}{v} + \frac{dp}{p} = 0,$$

whence $\qquad pv^\gamma = \text{constant}.$

15. *Isothermal expansion.* This is an expansion in which
the substance is kept at a constant temperature.

Since $pv = Rt$ for a perfect gas, $pv = \text{constant}$ in an
isothermal change.

16. *Isothermals and adiabatics.* Curves on the p-v diagram corresponding to isothermal and adiabatic changes
of a substance are called isothermals and adiabatics.

For a perfect gas they are the curves $pv = \text{constant}$ and
$pv^\gamma = \text{constant}$.

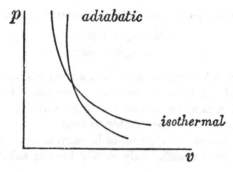

Since $\frac{dp}{dv} = -\frac{\gamma p}{v}$ for an adiabatic and $= -\frac{p}{v}$ for an isothermal, and $\gamma > 1$, an adiabatic crosses an isothermal as shown in the figure.

This property is true for any simple substance. (§ 74.)

17. Cycle. If a substance passes through a complete cycle of operations so as to return to the same state at the end of the cycle as at the beginning, the indicator diagram is a closed curve.

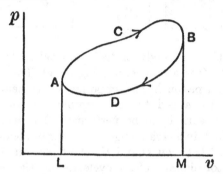

The work done by the substance in expanding along $ACB =$ area $ACBML$ and the work done on it in contracting along $BDA =$ area $BDALM$,

∴ the work done by it during the cycle
    = area $ACBML -$ area $BDALM$,
    = area of the cycle.

Again, since the substance is in the same state at the beginning and end of the cycle, the internal energy is unchanged. Therefore the heat taken in is equal to the work done by the substance.

Thus the area of the cycle also represents the heat taken in during the cycle.

18. Carnot's cycle. The following is an account of Carnot's cycle, as modified by Thomson so as to bring it into line with the dynamical theory of heat.

The system consists of a cylinder whose walls are impermeable to heat, but whose base D is a perfect conductor of heat. A and B are bodies whose temperature is kept constant, A the hot body or 'source' being at temperature

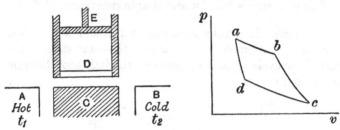

t_1 and B the cold body or 'refrigerator' at temperature t_2, $(t_1 > t_2)$. C is a non-conducting stand. The cylinder is fitted with a piston E, also impermeable to heat, which is connected by a rod to the external system on which mechanical work is to be performed. The working substance (which in a steam-engine is water or steam or both) is contained between D and E. The figure on the right is the indicator diagram of the cycle and consists of parts of two isothermals ab, cd corresponding to the temperatures t_1, t_2 and parts of two adiabatics ad, bc. There are four stages corresponding to the parts ab, bc, cd, da of the cycle.

(1) The substance is at temperature t_1 and its state is represented by the point a. The cylinder being placed on the source A, let the piston rise so that the substance takes in heat from A at constant temperature t_1, thus passing along the isothermal ab to the state b.

(2) The cylinder being removed and placed on the impermeable stand C, let the piston rise still further until the temperature has fallen to t_2, that of the refrigerator B. The substance has passed along the adiabatic bc to the state c.

(3) The cylinder being removed from C and placed on B, let the piston be pushed down so that the substance gives out heat to B at constant temperature t_2, thus passing along the isothermal cd to the state d.

(4) The cylinder being removed from B and placed on the impermeable stand C, the piston is still further pushed down until the temperature has risen to t_1, that of the source A. The substance has passed along the adiabatic da to its original state a.

The mechanical work done (W) in the cycle is equal to the area $abcda$. If Q_1 is the heat taken in at temperature t_1 along ab and Q_2 is the heat given out at temperature t_2 along cd, then since there is no transfer of heat to or from the substance along bc or da, $Q_1 - Q_2 = W$, units of work being used.

The efficiency of the cycle is the ratio of the mechanical work done to the heat taken in from the source, or W/Q_1, i.e. $\dfrac{Q_1 - Q_2}{Q_1}$.

19. *Reversibility of Carnot's cycle.* The cycle is reversible, that is, the substance can be taken along the path $adcba$ so that the stages 1, 2, 3, 4 are performed in the order 4, 3, 2, 1, each stage being carried out in the opposite sense to that just described.

The steps now are:

(4 reversed) The substance is in the state a at temperature t_1 and passes along the adiabatic ad to the state d, the temperature falling to t_2.

(3 reversed) It expands isothermally at temperature t_2 along dc to the state c, taking in an amount of heat Q_2 equal to that given out in stage 3 of the direct cycle.

(2 reversed) It passes adiabatically along cb to b, the temperature rising to t_1.

(1 reversed) It contracts isothermally along ba to the original state a at temperature t_1, giving out heat Q_1 equal to that taken in in stage 1 of the direct cycle.

The same work W is done, but is done *on* the substance by the piston, so that a quantity Q_2 of heat is taken from the cold body and a larger quantity Q_1 given out to the hot body, where $Q_1 - Q_2 = W$.

The reversibility of the operations depends on their being carried out sufficiently slowly for the substance at each stage to be virtually in thermal and mechanical equilibrium with its surroundings, *i.e.* the states through which it passes are a succession of equilibrium positions through which it is guided.

The temperature of A when heat is being taken in (during the direct cycle) by the substance is only infinitesimally greater than that of the substance—just enough for conduction to occur—and that of B when heat is being given out only infinitesimally less than that of the substance.

Also, the external pressure on the piston differs from that of the substance by an infinitesimal amount throughout—just enough to move the piston in or out as the case may be.

Such operations are reversible in every detail—for example, in the isothermal process represented by ab, at any stage a small outward motion of the piston will at first cause a small lowering of temperature which will suffice for heat to flow into the substance and maintain the level of temperature, whereas a small inward motion of the piston would cause a small rise of temperature sufficient to cause heat to flow out of the substance and keep the temperature constant. The slightest reversal of the motion of the piston would lead to a reversal of the flow of heat.

It is this reversibility of operations conducted so slowly that they may be regarded as a succession of equilibrium positions which has made Carnot's idea of a reversible cycle so fruitful in the study of chemical and physical changes.

20. *The second law of thermodynamics.* Carnot's principle that a reversible engine can produce the maximum amount of mechanical work derivable from a given quantity of heat let down through a given range of temperature has already been referred to. His argument depended upon the assumption that the heat taken from the hot body was wholly given up to the cold one. This assumption, that $Q_1 = Q_2$, must be replaced by $Q_1 - Q_2 = W$.

Clausius and Thomson each showed that Carnot's principle still holds if a new axiom, now known as the second law of thermodynamics, is used.

This axiom was given by Clausius as follows: 'It is impossible for a self-acting machine, unaided by any external agency, to convey heat from one body to another at a higher temperature.'

21. *Proof of Carnot's principle.* The principle is 'No heat engine working between two given temperatures of source and refrigerator can be more efficient than a reversible one.'

Let R be a reversible engine and S a non-reversible one.

Let R take in heat Q from the source and perform work W in its cycle, so that heat $Q - W$ is given out to the refrigerator. Suppose that in order to perform the same work W in its cycle, S takes in heat Q' from the source and gives out heat $Q' - W$ to the refrigerator. The respective efficiencies are $\dfrac{W}{Q}$, $\dfrac{W}{Q'}$. Let the non-reversible engine S be more efficient than the reversible one R. Then

$$\frac{W}{Q'} > \frac{W}{Q} \text{ or } Q > Q'.$$

Now let S drive R reversed, so that the output of work W from S is used up in driving R and the total work done in a cycle is zero.

S takes in heat Q' from the source and gives out heat $Q' - W$ to the refrigerator; R reversed takes in heat $Q - W$ from the refrigerator and gives out heat Q to the source.

Hence the refrigerator parts with heat

$$(Q - W) - (Q' - W)$$

and the source receives heat $Q - Q'$ with no output or

reception of external work. Thus heat $Q - Q'$ (which is positive, since $Q > Q'$) is transferred from the refrigerator to the source by a self-acting machine on which no external work is done. This is contrary to the axiom of Clausius (the second law). Hence S cannot be more efficient than R. This is Carnot's principle.

22. *All reversible engines working between the same two temperatures are of equal efficiency.* Consider two reversible engines R, R'. By using R to drive R' reversed it can be shown, as above, that R is not more efficient than R'; and, by using R' to drive R reversed, that R' is not more efficient than R. Hence R, R' are equally efficient.

23. *The theory of a reversible engine working between two given temperatures.* Since the efficiency of every reversible engine taking in heat Q_1 at temperature t_1 and giving out heat Q_2 at temperature t_2 is the same, $\dfrac{Q_1 - Q_2}{Q_1}$ or $1 - \dfrac{Q_2}{Q_1}$ is a function of t_1, t_2 only; *i.e.* $\dfrac{Q_1}{Q_2} = f(t_1, t_2)$.

Now consider two such engines, the first of which takes in heat Q_1 at temperature t_1 and gives out heat Q_2 at temperature t_2, the rejected heat Q_2 being taken in at that temperature by the second engine, which rejects heat Q_3 at temperature t_3. Then

$$\frac{Q_1}{Q_2} = f(t_1, t_2), \quad \frac{Q_2}{Q_3} = f(t_2, t_3).$$

Since the effect is that Q_1 is taken in at temperature t_1 and Q_3 given out at temperature t_3 by a reversible process (the combined action of two reversible engines),

$$\frac{Q_1}{Q_3} = f(t_1, t_3).$$

$$\therefore \ f(t_1, t_2) \cdot f(t_2, t_3) = f(t_1, t_3),$$

or
$$f(t_1, t_2) = \frac{f(t_1, t_3)}{f(t_2, t_3)}.$$

Now let t_3 be a constant standard temperature, while t_1, t_2 are variable. Then t_3 may be omitted from $f(t_1, t_3)$ and $f(t_2, t_3)$, so that they may be written $\psi(t_1)$ and $\psi(t_2)$. Hence

$$f(t_1, t_2) = \frac{\psi(t_1)}{\psi(t_2)},$$

$$\therefore \frac{Q_1}{Q_2} = \frac{\psi(t_1)}{\psi(t_2)}.$$

24. *Thomson's absolute thermodynamical scale of temperature.* Thomson[1] perceived that this property of a reversible engine could be used to compare the two temperatures t_1, t_2, the scale of temperature depending on the form of the function ψ chosen; and, that any such scale would be independent of the nature of any particular substance, as this property of a reversible engine did not depend upon the nature of the working substance. He finally chose a scale for which the function $\psi(t)$ was proportional to t so that $\dfrac{Q_1}{Q_2} = \dfrac{t_1}{t_2}$, and in his memoir on 'The dynamical theory of heat'[2] defined absolute temperature in the words 'the absolute values of two temperatures are to one another in the proportion of the heat taken in to the heat rejected in a perfect thermodynamic engine working with a source and refrigerator at the higher and lower of the two temperatures respectively.'

A thermodynamic thermometer would consist of a set of reversible engines each doing the same amount of work W in the cycle. The first takes in heat Q_1 at temperature t_1 and rejects heat Q_2 at temperature t_2; the second takes in the heat Q_2 rejected by the first at temperature t_2 and gives out heat Q_3 at temperature t_3 and so on.

Now $W = Q_1 - Q_2 = Q_2 - Q_3 = \dots$ etc.,

and $\dfrac{Q_1}{t_1} = \dfrac{Q_2}{t_2} = \dfrac{Q_3}{t_3} = \dots$ etc.,

$\therefore\ t_1 - t_2 = t_2 - t_3 = \dots$ etc.

---

[1] *Proc. Camb. Phil. Soc.* June 1848.   [2] *Coll. Papers,* vol. I. p. 235.

In this way equal temperature intervals on the absolute scale are indicated. The quantities of heat Q_1, Q_2, ... passed on from engine to engine diminish in proportion to the falling temperatures t_1, t_2, ..., so that when $t = 0$ no heat can be passed on and the substance is at the lowest possible limit of temperature. Thus the two laws of thermodynamics lead to the conclusion that the zero of the absolute thermodynamic scale is the lowest attainable temperature.

25. *The efficiency of a reversible engine working between two given temperatures.* Let the temperatures of the source and refrigerator be t_1, t_2.

The efficiency is $\dfrac{Q_1 - Q_2}{Q_1}$, where $\dfrac{Q_1}{t_1} = \dfrac{Q_2}{t_2}$.

$\therefore$ the efficiency is $\dfrac{t_1 - t_2}{t_1}$ or $1 - \dfrac{t_2}{t_1}$.

This, then, is the maximum efficiency attainable by any engine which takes in heat at temperature t_1 and rejects heat at temperature t_2.

The maximum work obtainable from a quantity of heat Q_1 taken in by the engine is $Q_1\left(1 - \dfrac{t_2}{t_1}\right)$.

Thus if a quantity of heat Q is available at temperature t and the lowest available temperature is t_0, the maximum mechanical work obtainable from it is $Q\left(1 - \dfrac{t_0}{t}\right)$.

This celebrated formula, due to the combined work of Carnot, Joule, Clausius and Thomson, has since their time become the foundation of the many and important advances in physical and chemical theory due to the use of thermodynamical principles.

26. *Carnot's cycle for a perfect gas.* For a perfect gas $pv = Rt$, where t is the absolute temperature on the perfect gas thermometer scale.

The isothermals ab, cd are $pv = Rt_1$, $pv = Rt_2$.

The adiabatics bc, ad are $pv^\gamma = \kappa_1$, $pv^\gamma = \kappa_2$.

By the first law, $dQ = dE + p\,dv$. As the energy E of a

perfect gas is a function of t only, E is constant along ab, so that $dE = 0$. $\therefore$ Along ab, $dQ = p\,dv$.

$\therefore$ Along ab the heat taken in,

$$Q_1 = \Sigma dQ = \int_{v_a}^{v_b} p\,dv = Rt_1 \int_{v_a}^{v_b} \frac{dv}{v} = Rt_1 \log\left(\frac{v_b}{v_a}\right).$$

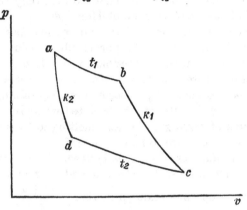

So, the heat given out along cd,

$$Q_2 = Rt_2 \log\left(\frac{v_c}{v_d}\right).$$

Now at a, $pv = Rt_1$, $pv^\gamma = \kappa_2$,

$$\therefore \ v_a{}^{\gamma-1} = \frac{\kappa_2}{Rt_1};$$

and at b, $pv = Rt_1$, $pv^\gamma = \kappa_1$,

$$\therefore \ v_b{}^{\gamma-1} = \frac{\kappa_1}{Rt_1},$$

$$\therefore \ \frac{v_a}{v_b} = \left(\frac{\kappa_2}{\kappa_1}\right)^{\frac{1}{\gamma-1}}.$$

So

$$\frac{v_d}{v_c} = \left(\frac{\kappa_2}{\kappa_1}\right)^{\frac{1}{\gamma-1}},$$

$$\therefore \ \frac{v_b}{v_a} = \frac{v_c}{v_d},$$

$$\therefore \ \frac{Q_1}{Q_2} = \frac{t_1}{t_2}.$$

This result shows that the ratio of two temperatures on the perfect gas thermometer is the same as their ratio on Thomson's absolute scale.

Since the permanent gases behave very approximately like a perfect gas, the ratio of two temperatures as recorded by an actual gas thermometer differs only slightly from that of the absolute scale of Thomson.

It will be shown in a later chapter how the experiments of Joule and Thomson on the porous plug cooling effect due to a given gas make it possible to calculate the ratio of two temperatures on Thomson's absolute scale when they have been observed by a thermometer containing that gas. In this way the ratio of the absolute thermodynamical temperatures of boiling water and melting ice was found at atmospheric pressure to be nearly 1·365[1].

Taking the number of degrees between these temperatures on the Thomson scale to be 100, and the temperature of melting ice on that scale as x, then $\dfrac{100 + x}{x} = 1\cdot365$, so that x is determined. Thomson found the value 273·7 for x, but more recent work gives the value 273·1. Thus 273·1 is the absolute thermodynamical temperature of melting ice and 373·1 that of boiling water at atmospheric pressure. Hence the absolute zero of the Thomson scale is 273·1 degrees below the melting point of ice, if 100 Thomson degrees is the range from the melting point of ice to the boiling point of water at atmospheric pressure.

The importance of the thermodynamical scale in low temperature work is clear, as a gas thermometer would behave anomalously at these temperatures; the temperature can only be found by thermodynamical calculations based on Thomson's scale.

---

[1] *Phil. Trans.* 1854.

CHAPTER III

DISSIPATION OF MECHANICAL ENERGY; ENTROPY

27. *The Carnot-Clausius equation for a reversible cycle.* Consider a system which goes through a reversible cycle of operations.

Divide the cycle into a large number (n) of parts in which heat Q_1 is taken in at temperature t_1, Q_2 at temperature t_2, and so on, respectively.

Consider $(n-1)$ reversible engines of which

the first takes in heat q_1 at temperature t_1
   and gives out $Q_2 + q_2$ at temperature t_2,

the second takes in heat q_2 at temperature t_2
   and gives out $Q_3 + q_3$ at temperature t_3,

and so on, as indicated in the diagram below:

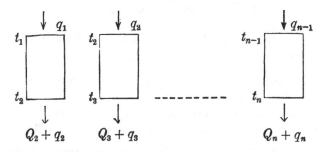

These engines, as a whole, take in q_1 at temperature t_1 and give out Q_2 at t_2, Q_3 at t_3, ..., Q_{n-1} at t_{n-1} and $(Q_n + q_n)$ at t_n.

The system considered takes in Q_1 at t_1, Q_2 at t_2, ..., Q_n at t_n.

The engines and the system form a complex system which is reversible; for this complex system, the net intake of heat is $Q_1 + q_1$ at temperature t_1 and the net output of heat is q_n at t_n.

$$\therefore \frac{Q_1 + q_1}{t_1} = \frac{q_n}{t_n}.$$

But for the separate engines,

$$\frac{Q_2 + q_2}{t_2} = \frac{q_1}{t_1},$$

$$\frac{Q_3 + q_3}{t_3} = \frac{q_2}{t_2},$$

$$\cdots\cdots\cdots\cdots\cdots$$

$$\frac{Q_n + q_n}{t_n} = \frac{q_{n-1}}{t_{n-1}}.$$

Adding all these equations we have

$$\frac{Q_1}{t_1} + \frac{Q_2}{t_2} + \dots + \frac{Q_n}{t_n} = 0,$$

or for infinitesimal changes, where dQ is the heat taken in at temperature t,

$$\Sigma \frac{dQ}{t} = 0, \text{ for a reversible cycle.}$$

This is the Carnot-Clausius equation and the proof in its essentials is that given by Thomson[1].

28. *Irreversible cycle.* If Q_1 is the heat taken in by an irreversible engine at temperature t_1 and Q_2 that rejected at temperature t_2, then since it is less efficient than a reversible one between those temperatures, its efficiency $\frac{Q_1 - Q_2}{Q_1}$ is less than $\frac{t_1 - t_2}{t_1}$ which is that of a reversible one.

$$\therefore 1 - \frac{Q_2}{Q_1} < 1 - \frac{t_2}{t_1},$$

or
$$Q_2 t_1 > Q_1 t_2,$$

or
$$\frac{Q_2}{t_2} > \frac{Q_1}{t_1}.$$

[1] *Coll. Papers*, vol. I. p. 236.

Consider now a system going through an irreversible cycle of operations.

Let it work in conjunction with $(n-1)$ reversible engines in the manner of the preceding paragraph. Using the result just proved, we shall have, since the complex system is irreversible,

$$\frac{Q_1 + q_1}{t_1} < \frac{q_n}{t_n},$$

while $\dfrac{Q_2 + q_2}{t_2} = \dfrac{q_1}{t_1}, \ldots, \dfrac{Q_n + q_n}{t_n} = \dfrac{q_{n-1}}{t_{n-1}}$, as before.

Adding, we have

$$\frac{Q_1}{t_1} + \ldots + \frac{Q_n}{t_n} < 0,$$

or $\Sigma \dfrac{dQ}{t} < 0$, for an irreversible cycle.

29. *The dissipation of mechanical energy in natural processes.* In any reversible process where heat is developed from mechanical energy a reversal of the process restores the mechanical energy from the heat; there is no dissipation (or waste) of *mechanical* energy, it can always be recovered by the system itself.

The spontaneous changes in nature are, however, irreversible; the initial state of a system cannot be restored from the final one by its own efforts, whatever human control of them there may be; to do this, mechanical energy from outside the system must be brought in, which means that in a natural change mechanical energy has been dissipated. (The energy itself is conserved, but the part of it available for mechanical work is reduced.)

Some of these irreversible processes will now be considered.

(i) *Production of heat by friction.* Joule determined the mechanical equivalent of heat by an experiment in which a falling weight was used to revolve a paddle in water so as to heat it by the friction between the paddle and water. The system at the end cannot be guided back to its original state, except by the use of fresh mechanical effort from outside. By falling, the weight has done mechanical work W and the water has received heat W. To restore the system to its original state, this heat W must be removed from the water and work W done on the weight to lift it back to its old level. If t is the temperature of the water, the maximum mechanical work obtainable from the heat W removed is $W\left(1 - \frac{t_0}{t}\right)$, where t_0 is the lowest available temperature, and this is not sufficient to raise the weight back to its old level. There is a deficiency of mechanical energy equal to $\frac{Wt_0}{t}$ which has been lost in the process and cannot be recovered.

(ii) *Conduction of heat.* Let a quantity of heat Q be transferred along a rod of metal by conduction from temperature t_1 at one end to temperature t_2 at the other $(t_1 > t_2)$.

If the lowest available temperature is t_0, the mechanical work obtainable from the heat before the transfer is

$$Q\left(1 - \frac{t_0}{t_1}\right) \text{ and after is } Q\left(1 - \frac{t_0}{t_2}\right).$$

There is a loss of mechanical energy equal to

$$Q\left(1 - \frac{t_0}{t_1}\right) - Q\left(1 - \frac{t_0}{t_2}\right) \text{ or } Qt_0\left(\frac{1}{t_2} - \frac{1}{t_1}\right),$$

which is positive, since $t_1 > t_2$.

(iii) *Gas rushing into a vacuum* (as in Joule's experiment). A gas under compression in a cylinder A is allowed to rush into a vacuous cylinder B so as to fill both vessels. If the gas be supposed perfect, the temperature and internal energy of the gas are unaltered. The initial state can be restored by using a piston in B to slowly push the gas back into A.

During this process the internal energy of the gas is unaltered (for a perfect gas it is independent of the volume), so that from the equation $dQ = dE + dW$, $dQ = dW$ or the mechanical work done is equal to the heat given to the gas.

Let each of these be W. To obtain the initial state, this heat W must be withdrawn; from this heat mechanical work $W\left(1 - \dfrac{t_0}{t}\right)$ can be obtained, where t is the temperature of the gas and t_0 the lowest available temperature. Thus the net expenditure of mechanical work in restoring the initial state is $W - W\left(1 - \dfrac{t_0}{t}\right)$ or $\dfrac{Wt_0}{t}$. This is the amount of mechanical energy wasted in the original process.

$\left[\text{If } v \text{ is the volume of } A \text{ and } v' \text{ that of } B,\right.$

$$W = \int_v^{v+v'} p\,dv = \int_v^{v+v'} \frac{Rt\,dv}{v} = Rt \log\left(\frac{v + v'}{v}\right)$$

and the dissipation is $\left.Rt_0 \log\left(\dfrac{v + v'}{v}\right).\right]$

Thomson was the first to perceive this tendency in nature of energy to become less and less available for the production of mechanical work, and formulated it in 1854[1] in the words 'there is at present in the material world a tendency to the dissipation of mechanical energy.'

Much later, in 1874[2], he gave a molecular theory of the dissipation of energy. If any man had under his orders a sufficient number of minute personages (the 'demons' of Clerk Maxwell)[3] who could guide the motion of each molecule of a gas, then it would be possible to reverse processes which to man alone are irreversible. The dissipation is due to man not being able to control the individual molecules but only to control them in large numbers, though 'in the phenomena of high vacua and radioactivity, where the molecules come nearly individually before our attention we get a partial glimpse of the motions contemplated by Maxwell and Thomson[4].'

[1] *Coll. Papers*, vol. I. p. 511.   [2] *Coll. Papers*, vol. V. p. 11.
[3] Letter from Clerk Maxwell to Lord Rayleigh—*Life of Lord Rayleigh, by his son*, p. 47.
[4] Sir J. Larmor's obituary notice of Lord Kelvin, *Proc. Roy. Soc.* 1908.

30. *Entropy.* While Thomson was developing the principle of the dissipation of mechanical energy, the same ideas were being carried through by Clausius (1855) by the use of the concept of 'entropy.'

If in a reversible change, a substance takes in heat dQ when its temperature is t, $\dfrac{dQ}{t}$ is called the increase of 'entropy' of the substance. Denoting this by $d\phi$, we have $dQ = t\,d\phi$.

If the substance passes from the state A to the state P by a reversible path AKP, the increase of entropy is

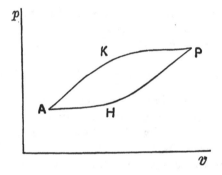

$\Sigma \dfrac{dQ}{t}$ along AKP; and if by the path AHP, the increase is $\Sigma \dfrac{dQ}{t}$ along AHP.

But $AHPK$ is a reversible cycle for which $\Sigma \dfrac{dQ}{t} = 0$,

$$\therefore \quad \Sigma \frac{dQ}{t} \text{ along } AKP + \Sigma \frac{dQ}{t} \text{ along } PHA = 0,$$

or $\qquad \Sigma \dfrac{dQ}{t}$ along $AKP = \Sigma \dfrac{dQ}{t}$ along AHP.

Thus the increase of entropy from A to P is independent of the path from A to P, or $\phi_P - \phi_A$ is independent of the path from A to P and depends only on the states A, P. If the state A is taken as the state of zero entropy, the entropy ϕ in any other state is $\Sigma \dfrac{dQ}{t}$ for any reversible path from A to that state.

31. *Increase of entropy in natural processes.* If a substance passes from one state to another by an irreversible process, the increase of entropy is $\Sigma \dfrac{dQ}{t}$ taken for any *reversible* process by which the substance can pass from the first state to the second.

For instance, in Joule's experiment, where a gas under compression in a cylinder A rushes into a vacuous cylinder B so as to fill both vessels.

Supposing the gas perfect, the temperature is unchanged; in the first state the gas has volume v, temperature t and in the final state has volume $(v + v')$, temperature t.

The change of state can be effected by the following reversible process.

Let the volume A of gas expand slowly so as to press out a piston in B so controlled that the external pressure is just less than that of the gas at all stages of the expansion and let heat flow into the gas through the walls of the vessels so as to keep the temperature constant. In this way the second state is reached.

The increase of entropy in the irreversible process is equal to $\Sigma \dfrac{dQ}{t}$ for this reversible one $= \dfrac{1}{t} \Sigma dQ$, since t is constant, $= \dfrac{Q}{t}$, where Q is the heat taken in. In the reversible process E is constant,

$$\therefore \ dQ = dW \ \text{ or } \ Q = W$$

and the increase of entropy $= \dfrac{W}{t}$, where W is the work done by the gas in the reversible expansion. This

$$= \frac{1}{t} \int_v^{v+v'} p\, dv = R \int_v^{v+v'} \frac{dv}{v} = R \log \left(\frac{v + v'}{v} \right),$$

where v, v' are the volumes of A, B.

It may be observed

(i) that though no external heat has been absorbed in the irreversible process there is an increase of entropy;

(ii) that Thomson's dissipation of energy in this experiment is t_0 times Clausius' increase of entropy (p. 29).

Again, in the irreversible process of conduction of heat where a quantity of heat Q is conducted from A where the temperature is t_1 to B where it is t_2 $(t_1 > t_2)$, A loses entropy $\dfrac{Q}{t_1}$, B gains $\dfrac{Q}{t_2}$ and the gain of entropy of the system is $Q\left(\dfrac{1}{t_2} - \dfrac{1}{t_1}\right)$, which is positive.

Thus it appears that in natural processes there is an increase of entropy, or as Clausius said 'the entropy of the universe tends towards a maximum.' Thomson's view that available energy is always lost, never gained, and Clausius' view that entropy is always gained, never lost, express the same natural principle.

32. *Irreversible engine.* It is of interest to consider an irreversible engine working between two given temperatures from the points of view of available energy and entropy.

(i) *Thomson.* If Q_1 is the heat taken in by the engine from the source at temperature t_1 and Q_2 is the heat given out to the refrigerator at temperature t_2, the efficiency $\dfrac{Q_1 - Q_2}{Q_1}$, being less than that of a reversible one, is $< \dfrac{t_1 - t_2}{t_1}$.

Hence
$$\frac{Q_2}{t_2} > \frac{Q_1}{t_1}.$$

The engine receives heat Q_1 at temperature t_1 and receives $- Q_2$ at temperature t_2 and does work

$$W = Q_1 - Q_2.$$

If the lowest available temperature is t_0, the energy

available for mechanical work received by the engine from the source and refrigerator is

$$Q_1 \left(1 - \frac{t_0}{t_1}\right) - Q_2 \left(1 - \frac{t_0}{t_2}\right) - W$$

or

$$t_0 \left(\frac{Q_2}{t_2} - \frac{Q_1}{t_1}\right);$$

which is positive.

But at the end of the cycle, the engine is in the same condition as at the beginning. Hence the irreversible processes occurring inside the engine (such as conduction of heat from one part to another, friction, and throttling) have dissipated $t_0 \left(\dfrac{Q_2}{t_2} - \dfrac{Q_1}{t_1}\right)$ of the available energy received from the source and refrigerator.

If the engine were reversible, $\dfrac{Q_1}{t_1} = \dfrac{Q_2}{t_2}$, and there would be no dissipation.

(ii) *Clausius*. The source loses entropy $\dfrac{Q_1}{t_1}$ and the refrigerator gains entropy $\dfrac{Q_2}{t_2}$ and the two gain entropy $\left(\dfrac{Q_2}{t_2} - \dfrac{Q_1}{t_1}\right)$. The engine having completed a cycle is in the same state as before and its entropy is unchanged. Thus the whole system at the end of a cycle has gained entropy $\left(\dfrac{Q_2}{t_2} - \dfrac{Q_1}{t_1}\right)$.

In this case, too, Thomson's dissipation of mechanical energy is equal to t_0 times Clausius' increase of entropy.

33. *Irreversible cycle.* These results are easily extended to the case of any irreversible cycle, where at any stage dQ is the heat taken in and t is the corresponding temperature.

The source and refrigerator lose entropy $\Sigma \dfrac{dQ}{t}$ in the cycle; the engine is in the same condition at the end of the cycle as before, so that its entropy is unchanged. Hence the gain of entropy is $-\Sigma \dfrac{dQ}{t}$, which is positive. [It has been shown that $\Sigma \dfrac{dQ}{t} < 0$ for an irreversible cycle.]

Again, the available energy received by the engine from the source and refrigerator is

$$\Sigma\, dQ \left(1 - \frac{t_0}{t}\right) - W,$$

where W is the work done in the cycle. The engine is in the same condition at the end of the cycle as before; hence this available energy has been dissipated in the engine.

Now $\Sigma dQ = W$,

$\therefore$ the dissipation of mechanical energy $= -\, t_0 \,\Sigma \dfrac{dQ}{t}.$

34. *Entropy and probability.* Reference has already been made to Thomson's molecular theory of the dissipation of energy (1874)[1]; in an appendix, he discusses the question of irreversibility from a new point of view, depending on the theory of probability.

He considers a vessel containing air, in which oxygen and nitrogen are uniformly diffused in the proportion of $1:4$. What is the probability that a given part of the vessel equal in volume to one-fifth of the whole shall at some later time contain all the oxygen and the remaining four-fifths all the nitrogen? *i.e.* what is the probability of the process of diffusion of the two gases being completely reversed? The result found was

$$p = (\tfrac{1}{5})^n\, (\tfrac{4}{5})^{4n}$$

where p is the probability, n the number of molecules of oxygen and $4n$ the number of molecules of nitrogen.

If there were 5 c.c. of air, $n = 2{\cdot}7 \times 10^{19}$, the number of molecules of a gas per c.c.

---

[1] *Coll. Papers*, voL v. p. 11.

Now

$$\log p = 8n \log 2 - 5n \log 5 = n \, (13 \log 2 - 5),$$
$$= (- 1{\cdot}08661) \, n,$$
$$\therefore \quad p = 10^{-(1{\cdot}08661)n},$$
$$= 10^{-28338 \times 10^{15}},$$

or $p = 10^{-N}$ where N is a number containing 20 figures; and the index N would increase in proportion to the volume of air taken.

This result means that the probability of the diffusion being reversed is so small that it will never be observed; diffusion is irreversible.

35. This new outlook on thermodynamics was developed by Boltzmann in 1877[1]. Taking the principle of Clausius that a system tends to pass towards states of greater entropy to be also the principle that a system of molecules tends to pass towards states of greater probability of occurrence, he showed that the entropy of a given state of a system of molecules is proportional to the logarithm of the probability of its occurrence.

This leads to the formula

$$\frac{1}{Rt} = \frac{d}{dE} \, (\log W),$$

where W is the probability that the energy of such a system has the value E. This is the equivalent of $dQ = t \, d\phi$, with $R \log W$ as the entropy ϕ. The radiation formula of Planck[2], that the mean energy of a vibrator of frequency ν is equal to

$$Rt \, \frac{x}{e^x - 1}, \text{ where } x = \frac{h\nu}{Rt},$$

was deduced by the use of this equation and his now famous concept of the 'quantum' of energy $h\nu$.

[1] *Vorlesungen über Gas-theorie*, p. 42.
[2] *La Théorie du Rayonnement et les Quanta*, Gauthier-Villars, 1912, p. 104.

36. *Adiabatic change.* In an adiabatic change the process is reversible and no heat enters or leaves the system. Hence $dQ = td\phi$ and $dQ = 0$, $\therefore\ d\phi = 0$ and ϕ is constant during the change.

The adiabatics are curves of constant entropy.

37. *Entropy-temperature diagram* (Willard Gibbs). The state of the substance is represented by a point whose co-

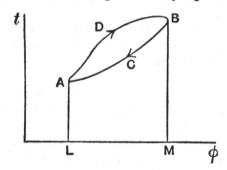

ordinates are $(\phi,\ t)$. The isothermals are lines parallel to the ϕ axis and the adiabatics are lines parallel to the t axis.

Let the substance pass from A to B by the reversible path ADB.

The heat taken in is

$$\Sigma dQ = \Sigma t d\phi = \int_A^B t d\phi = \text{area } ADBML,$$

or the area 'under' ADB.

So in passing from B to A by the reversible path BCA, the heat given out is equal to the area under ACB.

Hence in the cycle $ADBCA$, the heat taken in is equal to the difference of these areas and is equal to the area of the diagram of the cycle.

Now by the first law, $dQ = dE + dW$ and after a cycle, E is unchanged so that $dE = 0$, $\therefore\ dQ = dW$, or the heat taken in is equal to the mechanical work done. Hence the work done in the cycle is also equal to the area of the diagram of the cycle.

38. *The ϕ-t diagram for Carnot's cycle.* The isothermals are *ab*, *dc* and the adiabatics are *bc*, *ad*; the diagram of the cycle is a rectangle.

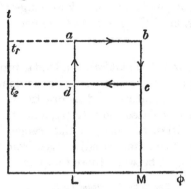

Q_1, the heat taken in from the source, is equal to the area under *ab*, Q_2, the heat given out to the refrigerator, is equal to the area under *dc*,

$$\therefore \; \frac{Q_1}{Q_2} = \frac{\text{area under } ab}{\text{area under } dc} = \frac{aL}{dL} = \frac{t_1}{t_2}.$$

The ϕ-t diagram was first given by Willard Gibbs in 1873[1] in a memoir on the use of graphical methods in thermodynamics; it has since found important applications in the study of the steam-engine, as Gibbs foresaw on account of the simple form which Carnot's cycle for a perfect engine takes on the diagram.

39. *Calculation of entropy.* Consider a perfect gas, for which $dQ = c_v dt + p dv$,

$$\therefore \; td\phi = c_v dt + p dv,$$

$$d\phi = c_v \frac{dt}{t} + \frac{p\,dv}{t}, \; \text{and} \; pv = Rt,$$

$$\therefore \; d\phi = c_v \frac{dt}{t} + R \frac{dv}{v},$$

$$\therefore \; \phi = c_v \log t + R \log v + C.$$

[1] *Trans. Connecticut Acad.* II.; *Coll. Papers*, vol. I. p. 9.

The value of the constant C depends on the volume and temperature of the gas when in the standard state of zero entropy.

This result depends upon the knowledge of the characteristic equation of the gas and of the constancy of c_v for the gas.

40. *Callendar's steam tables*[1]. Callendar proposed a characteristic equation to represent the behaviour of steam in the range of temperature used in practice (0° to 250° C.), and by its use calculated the entropy, and other thermodynamical quantities, of steam in that range. The entropy per pound of steam (saturated) ranges from 2·17 to 1·47 lb.-calories/deg. C. between those temperatures.

[1] *The Callendar Steam Tables.* 1915.

CHAPTER IV

THERMODYNAMICS OF A FLUID; CHANGE OF STATE

41. *Vapours.* AB is the column of mercury in a barometer with a vacuum above A. If a few drops of a volatile liquid, such as ether, are inserted at B so as to rise up into the vacuum, they are vaporised and the column is depressed to A', the pressure of the vapour being measured by the difference of level between A and A'. The introduction of more ether leads to further depression of A' until a stage is reached when a drop of ether on reaching the surface A' is no longer vaporised and remains liquid. The vapour in this state is called 'saturated vapour'; and the pressure of the vapour is called 'the saturation vapour pressure' or usually just 'vapour pressure.' Its value for a given fluid depends only upon the temperature at which the experiment is carried out; it is found to increase with the temperature.

In the earlier stages, where fresh liquid introduced was vaporised, the vapour was 'unsaturated.'

All gases are unsaturated vapours; such vapours approximately satisfy $pv = Rt$, and the more closely the further they are from the point of liquefaction; thus at constant temperature if the volume diminishes the pressure rises.

42. Vapours in contact with their liquid, and therefore saturated, have the property that at constant temperature, if the volume changes the pressure remains constant. This can be shown by the following experiment.

Fig. *A* represents a barometer tube containing a saturated vapour in contact with its liquid. The tube is held in a deep vessel. If the tube is now raised to the position in Fig. *B*, the level of the mercury column is unchanged, so long as there is liquid remaining in contact with the vapour. The volume of the vapour has increased, but its pressure is unaltered.

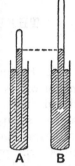

During the passage from *A* to *B* fresh liquid is vaporised to maintain the pressure constant; if the tube were lowered again to the position *A* this liquid would reappear by condensation from the vapour.

A   B

43. *Isothermals of a liquid and its vapour.* The figure represents a portion of an isothermal of a fluid. Consider

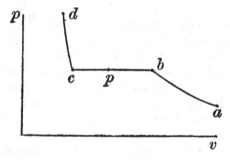

the fluid as an unsaturated vapour in the state *a*. Let the pressure be increased at constant temperature. The vapour contracts along the path *ab*, until at *b* it becomes saturated. This occurs at a definite pressure for a given fluid, the pressure depending only upon the temperature. Condensation then begins and the pressure remains constant with decreasing volume until the whole becomes liquid, at the state *c*. Along *bc* the substance is a mixture of saturated vapour and liquid in varying proportions. As the pressure rises there is only very slight compression of the liquid; this is shown by the path *cd*.

At any point p of bc the liquefaction is only partial; let x, $1 - x$ be the weights of liquid and vapour in unit mass of the mixture.

If v_p, v_b, v_c are the volumes of unit mass of the mixture, vapour, liquid, then
$$v_p = xv_c + (1 - x)\, v_b$$
or
$$x = \frac{v_b - v_p}{v_b - v_c} = \frac{pb}{bc}.$$

∴ Ratio of weights of liquid and vapour is $pb : pc$.

Thus the point p indicates the pressure, volume, and *composition* of the mixture.

44. *Continuity of the liquid and gaseous states of a fluid.* The isothermals of carbonic acid were first mapped out by Andrews[1] for a considerable range of temperature and pressure. They are of the form shown in the figure.

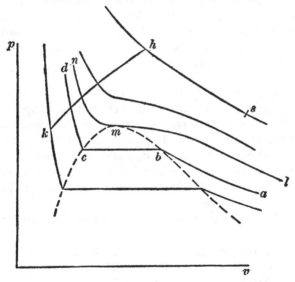

$abcd$ is an isothermal of the type considered above (§ 43).

The straight part bc, which corresponds to states of condensation where the fluid is a mixture of liquid and vapour, is seen to shorten as the temperature rises, and for the isothermal lmn vanishes. The vapour becomes liquid at m without condensation; so that at m the liquid and vapour

[1] *Phil. Trans.* 1869.

are physically indistinguishable. The point *m* was called
by Andrews the 'critical point,' to which correspond a
'critical' temperature, volume, and pressure. Experimentally the critical point is found when the fluid, partly vapour
and partly liquid, suddenly becomes uniform in appear-
ance without any visible surface of separation.

The dotted curve divides the figure into two regions; in
the region inside, the fluid can exist as a mixture of liquid
and vapour; in the region outside only either as liquid or
as vapour. Thus in the passage by any path between two
states, condensation or vaporisation can only occur if the
path crosses the region within the dotted curve. A path
such as *hk*, which does not fulfil this condition, would take
the fluid from the gaseous state at *h* to the liquid state at
k by a continuous process throughout which the fluid
remains homogeneous.

45. *Liquefaction of gases.* If a gas is compressed isothermally at a temperature above the critical temperature
(which corresponds to the isothermal *lmn*) the path would
be of the type *hs*, which does not cross the region within
the dotted curve, and no condensation can occur.

Thus a gas can only be liquefied by pressure if it is first
cooled below the critical temperature, a property predicted
by Faraday in 1826.

The following are some critical temperatures, and pressures:

Carbonic acid	31° C.	72 atmospheres
Oxygen	− 119° C.	58 ,,
Nitrogen	− 149° C.	28 ,,
Hydrogen	− 238° C.	15 ,,
Helium	− 264° C.	

46. *Ideal isothermal for a fluid.* Let *abcd* be a normal
isothermal for water and steam, for example.

It is possible to compress steam beyond the state of
saturation without condensation occurring, so that the

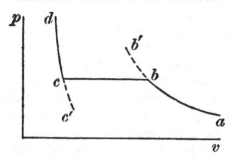

path *bb'* is described, if dust nuclei which promote condensation are removed.

It is also possible with care to heat water to a temperature of 105° C. at atmospheric pressure without vaporisation, which is equivalent to keeping it at a lower pressure than that at which it would normally vaporise at that temperature, so that the path *cc'* is described.

Both of these states are highly unstable.

James Thomson suggested that the curves *bb'*, *cc'* might, if prolonged, turn round and join, forming an ideal isothermal *abfgecd*.

The portions *bf*, *ce* correspond to real changes of a very unstable kind; the portion *ef* would require an increase of

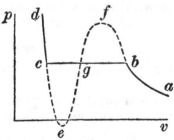

pressure to produce an increase of volume, which has not been realised for any fluid, as such a change would be completely unstable. But if such a continuous ideal isothermal is drawn for a fluid below its critical temperature, the actual one *abcd* can be found by drawing the line *bc* at such a level of pressure that the areas *ceg* and *gfb* are equal. Maxwell showed this by imagining the fluid taken round the isothermal cycle *cegfbgc*. Since $dQ = dE + dW$ and $dQ = td\phi$, we have $td\phi = dE + dW$. ∴ since t is constant,

t (change of entropy) = change of E + work done.

At the end of the cycle, the entropy and E are unchanged. Therefore, the work done is zero, or the area of the cycle is zero, *i.e.* the positive area *gfb* is equal in magnitude to the negative area *ceg*.

By this construction, if a set of ideal isothermals is known, the saturation pressure for each one can be found.

Many forms of characteristic equation for a fluid which gives such a set of ideal isothermals have been proposed. The first was given by Van der Waals.

47. *Van der Waals' equation.* This is a characteristic equation proposed by Van der Waals in 1873[1] for a fluid under all conditions ranging from the gaseous to the liquid state. It accounts, in a remarkable degree, for many of the most important properties of fluids.

The equation is $\left(p + \dfrac{a}{v^2}\right)(v - b) = Rt$, where a, b, R are constants for a given fluid, and was based on the kinetic theory of gases. The constant b cannot be much different from the liquid volume; the term $\dfrac{a}{v^2}$ represents the effect on the pressure of molecular attractions resisting the expansion of the gas[2].

If the ordinate y is equal to the pressure p and the abscissa x is equal to the volume v, the isothermals are the curves $\left(y + \dfrac{a}{x^2}\right)(x - b) = k$, where k is constant for a given isothermal; or $y = \dfrac{k}{x - b} - \dfrac{a}{x^2}$.

---

[1] *Phys. Memoirs of London Phys. Soc.* vol. I. pt. III. (Translation.)
[2] Jeans' *Dynamical Theory of Gases*, chap. vi.

The portions of these curves for $x > b$ are shown in the figure.

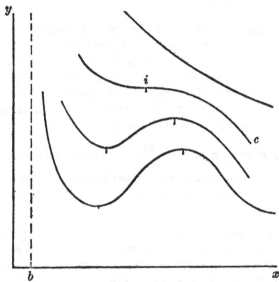

Each has a maximum and a minimum, whose abscissas approach as k increases until for the curve c they coincide at a point of inflexion i on the curve c. The curves below c are the ideal isothermals of James Thomson and i represents Andrews' critical state. The critical state then is given by

$$\frac{dy}{dx} = 0, \quad \frac{d^2y}{dx^2} = 0,$$

or

$$\left.\begin{array}{c} \dfrac{k}{(x-b)^2} = \dfrac{2a}{x^3} \\[2mm] \dfrac{k}{(x-b)^3} = \dfrac{3a}{x^4} \end{array}\right\}.$$

From these and $y = \dfrac{k}{x-b} - \dfrac{a}{x^2}$, it follows that

$$x = 3b, \quad k = \frac{8a}{27b}, \quad y = \frac{a}{27b^2},$$

or if p_0, v_0, t_0 are the critical pressure, volume and temperature,

$$p_0 = \frac{a}{27b^2}, \quad v_0 = 3b, \quad Rt_0 = \frac{8a}{27b}.$$

From the experiments of Regnault, the constants a, b, R were calculated for carbonic acid; the critical temperature and pressure were found to be 32° C. and 61 atm., the actual values found by experiment being 31° C. and 72 atm. In some other respects the agreement was not so good. For instance, experiments show that v_0 is about four times the liquid volume, or $v_0 = 4b$.

Also from the above, $\dfrac{Rt_0}{p_0 v_0} = \frac{8}{3} = 2 \cdot 67$, indicating how far from the state of a perfect gas the fluid is in the critical state; but experiment gives the value $3 \cdot 75$ for this ratio (S. Young[1]).

48. *The reduced equation.* If the critical pressure, volume and temperature are taken as units, and p', v', t' are any pressure, volume and temperature in terms of those units, $p = p_0 p'$, $v = v_0 v'$, $t = t_0 t'$.

Van der Waals' equation becomes

$$p_0 p' = \frac{Rt_0 t'}{v_0 v' - b} - \frac{a}{v_0^2 v'^2},$$

$$p' \frac{a}{27b^2} = \frac{t' 8a}{27b^2 (3v' - 1)} - \frac{a}{9b^2 v'^2},$$

$$p' = \frac{8t'}{3v' - 1} - \frac{3}{v'^2},$$

$$\left(p' + \frac{3}{v'^2}\right)(v' - \tfrac{1}{3}) = \tfrac{8}{3}t',$$

or dropping the dashes

$$\left(p + \frac{3}{v^2}\right)(v - \tfrac{1}{3}) = \tfrac{8}{3}t,$$

where the units of measurement of p, v, t are their critical values. This is the 'reduced' form of the equation and the values of p, v, t are 'reduced' values of the pressure,

[1] *Phil. Mag.* 1892.

volume, and temperature. The constants a, b, R for a particular fluid have disappeared, and the equation is the same for all fluids.

Thus if two gases have the same reduced pressure and volume, they have the same reduced temperature.

49. *Boiling point and pressure.* The figure shows an isothermal for temperature t; p, v, t are reduced values. This

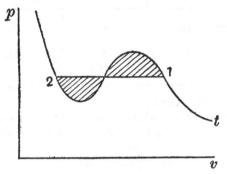

curve is the same for all gases, its form depending only upon t. If the line 12 is drawn parallel to the v axis to make the shaded areas equal, the positions of the points 1, 2 will also depend only upon t.

But 2 represents the liquid when on the point of boiling. Hence the reduced pressure at the boiling point depends only upon t or, at equal reduced pressures, the reduced boiling points are the same for all fluids.

This theoretical result was tested by S. Young[1] who found that for the halogen derivatives of benzene the agreement was good, though there were anomalies in the behaviour of some of the other fluids for which no 'reduced' equation could account.

50. *Amagat's experiments*[2]. Amagat compressed gases up to very high pressures and expressed his results by curves of which the ordinate y was pv, and the abscissa x was p.

---

[1] *Phil. Mag.* 1892.　　[2] *Ann. Chim. Phys.* 1881 and *C.R.* 1888.

For a perfect gas, the isothermals pv = constant would be parallel lines y = constant.

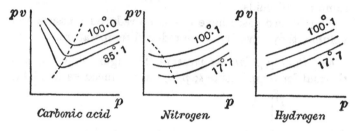

Carbonic acid   *Nitrogen*   *Hydrogen*

For carbonic acid, nitrogen, and hydrogen Amagat found curves of the form shown in the figures. In the case of carbonic acid and nitrogen, the curves have minima which lie on the dotted curves.

Van der Waals' equation accounts for these minima and gives a form for the dotted curve. Using the reduced form

$$\left(p + \frac{3}{v^2}\right)\left(v - \frac{1}{3}\right) = \kappa$$

for an isothermal, where $\kappa = \frac{8}{3}t$, and writing $y = pv$, $x = p$, the equation of an isothermal for an actual gas is

$$\left(x + \frac{3x^2}{y^2}\right)\left(\frac{y}{x} - \frac{1}{3}\right) = \kappa$$

or

$$\left(1 + \frac{3x}{y^2}\right)\left(y - \frac{x}{3}\right) = \kappa.$$

To find the minimum point, we have

$$3\left(\frac{1}{y^2} - \frac{2x}{y^3}\frac{dy}{dx}\right)\left(y - \frac{x}{3}\right) + \left(1 + \frac{3x}{y^2}\right)\left(\frac{dy}{dx} - \frac{1}{3}\right) = 0.$$

At a minimum point, $dy/dx = 0$, or

$$\frac{3}{y^2}\left(y - \frac{x}{3}\right) - \frac{1}{3}\left(1 + \frac{3x}{y^2}\right) = 0.$$

This, being independent of κ, is the curve on which the minima for the isothermals lie.

It reduces to $(9 - y)\,y = 6x$, a parabola, shown as the curve *Obac* on the next page.

Thus the dotted curve found for carbonic acid would correspond to part of Ob and that for nitrogen to part of ab.

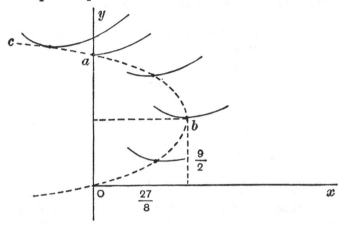

Now consider the isothermal through a. For this point $x = 0$, $y = 9$, therefore

$$\kappa = 9, \text{ or } \frac{8t}{3} = 9 \text{ or } t = \frac{27}{8} = 3\cdot375.$$

Thus for temperatures above 3·375 of the critical temperature, the minima are along the part ac, so that there are no minima for positive values of x, that is of p. This is shown by Amagat's diagram for hydrogen, where the temperatures ranged from 17° to 100° C.; for 3·375 of the critical temperature for hydrogen is $3\cdot375 \times 35$ abs. $= 118$ abs. $= -155°$ C., and Amagat's temperatures are high above this.

In the case of nitrogen however, 3·375 of the critical temperature is $3\cdot375 \times 124$ abs. $= 418$ abs. $= 145°$ C. and Amagat's temperatures, also 17° to 100° C., are all below this, so that the minima appear on the right of the axis of y.

The experiments of Witkowski[1] on air at temperatures ranging from $-145°$ to 100° C. showed the almost complete curve Oba of minima for the isothermals.

[1] *Phil. Mag.* 1896.

The theoretical value 3·375 for the reduced temperature of the isothermal passing through a is of the same order as the value found by experiment, which is 3.

51. *Other forms of characteristic equation.* Other forms of equation such as

$$p = \frac{Rt}{v-b} - \frac{c}{t(v+a)^2} \text{ (Clausius)[1]};$$

$$p = \frac{Rt}{v-b} e^{-\frac{a}{Rtv}} \text{ (Dieterici)[2]};$$

$$p = \frac{Rt}{v-b} - \frac{ap^{\frac{1}{3}}}{tv^{\frac{5}{3}}} \text{ (Lees)[3]}$$

have since then been proposed, to give closer quantitative results than that of Van der Waals. While Van der Waals' equation gives the values 31° C., 61 atm. for the critical temperature and pressure of carbonic acid, that of Clausius gives 31° C., 77 atm.; the experimental result is 32°, 72 atm. Again the ratio Rt/pv at the critical point is found by experiment to be 3·75; Van der Waals' equation gives the result 2·67, but Dieterici's equation the result $\frac{1}{2}e^2$, (where e is the base of logarithms), or 3·69. On the other hand in the theory of Amagat's experiments, experiment gives the temperature corresponding to the point called a as 3; Van der Waals' equation gives 3·375, whilst that of Dieterici gives 4.

Lees, starting with the form

$$p = \frac{Rt}{v-b} - \phi(p, v, t),$$

chooses the form of the function ϕ so that the equation agrees with certain known experimental results; the final form is that given above. It gives the ratio Rt/pv at the critical point as 3·75 exactly and for the Amagat point a the temperature 3·11.

[1] *Phil. Mag.* 1880.   [2] *Ann. der Physik*, 1901.   [3] *Phil. Mag.* 1924.

In addition to these equations which apply to all values of the variables, there are others, such as that of Callendar for steam and that of Berthelot for low pressures, which are intended to apply only in a restricted range of the variables. Callendar's equation

$$v = \frac{Rt}{p} + b - \frac{c}{t^n},$$

has been of the greatest value in compiling tables of the properties of steam within the range of temperatures usually employed in steam-engines, the value of n used for steam being $\frac{10}{3}$.

CHAPTER V

THERMODYNAMIC FUNCTIONS

52. *Internal Energy* (*E*).

Since $dE = dQ - dW$, the increase of E is equal to (the heat taken in by the substance) − (the work done by the substance).

Consider, for instance, 1 lb. of water at 0° C. under a pressure of 300 lb. per sq. in. and suppose it heated until it becomes saturated steam at this pressure. (The boiling point for this pressure is known to be 214°·32 C.) It is proposed to calculate the internal energy changes in this process.

(i) *Change from water at* 0° *to water at* 214°·32.

$$dE = dQ - pdv.$$

dQ is here = 218·2 calories, from tables, (the specific heat of water increases slightly above 1 as the temperature rises).

$$pdv = \frac{(300 \times 144)\,(\cdot 01898 - \cdot 01602)}{1400} \text{ calories} = \cdot 09 \text{ calorie.}$$

(The increase of volume is found from tables of water, and 1400 ft.-lb. = 1 calorie.)

∴ The increase of E = 218·11 calories.

Thus nearly all the heat given goes to increase the internal energy of the water.

(ii) *Change from water at* 214°·32 *to saturated steam at* 214°·32.

$dv = 1\cdot 583 - \cdot 019 = 1\cdot 564$, from tables of water and steam.

$$pdv = \frac{300 \times 144 \times 1\cdot 564}{1400} = 47\cdot 23 \text{ calories.}$$

The latent heat at that temperature is 454·8; this is dQ.

∴ Increase of E = 406·57 calories.

∴ The total increase in (i) and (ii) is 624·68 calories.

(Taking $E = 0$ for water at $0°$ C., the tables give 624·67 as E for saturated steam at 300 lb. pressure.)

This illustrates how E for any state of a fluid can be calculated from the heat received and work done in passing from any standard state.

Also for a reversible change, $dQ = dE + pdv$, and $dQ = td\phi$.

$\therefore$ $dE = td\phi - pdv$, for a simple substance.

53. *Total Heat* (I).

This is defined by the equation

$$I = E + pv.$$

(i) $\qquad\qquad \therefore\ dI = dE + pdv + vdp$

$$= td\phi + vdp.$$

Thus at constant volume

$$dE = td\phi = dQ,$$

and at constant pressure

$$dI = td\phi = dQ.$$

$\therefore$ The increase of E is the heat taken in at constant *volume*; the increase of I is the heat taken in at constant *pressure*.

(ii) If a substance expands *adiabatically*, $dQ = 0$ and $dI = vdp$.

$$\therefore\ I_a - I_b = \int_b^a vdp = \text{area } abnm.$$

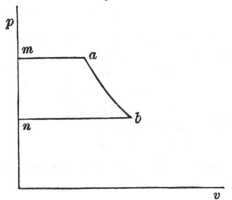

$I_a - I_b$ is called in engine theory the 'heat drop' in adiabatic expansion.

(iii) If a gas rushes through a small orifice from a higher pressure to a lower, the sides of the containing vessels or tubes being non-conductors of heat, I is unaltered.

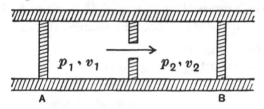

Let the piston A push a volume v_1 of gas at pressure p_1 through the orifice, which becomes of volume v_2 at pressure p_2 after passage $(p_1 > p_2)$.

The work done by A on the gas is $p_1 v_1$ and the work done on B by the gas is $p_2 v_2$.

$$\therefore\ dW = p_2 v_2 - p_1 v_1.$$

Now $dQ = dE + dW$, and $dQ = 0$ since no heat passes through the walls of the tube or the pistons.

$$\therefore\ dE + dW = 0.$$
$$\therefore\ (E_2 - E_1) + (p_2 v_2 - p_1 v_1) = 0.$$
$$\therefore\ E_2 + p_2 v_2 = E_1 + p_1 v_1$$

or $$I_2 = I_1.$$

Such a process is called by engineers a 'throttling' process.

54. *Thermodynamic potential at constant volume* (ψ).
This is defined by the equation
$$\psi = E - t\phi.$$
$$\therefore\ d\psi = dE - td\phi - \phi dt$$
$$= -pdv - \phi dt.$$
Thus at constant temperature
$$- d\psi = pdv,$$
or the decrease of ψ is equal to the work done by the substance.

Hence ψ for any state of the substance is the work done on the substance in passing at constant temperature t from a standard state (for which ψ is zero) to the actual one; conversely, ψ measures the energy available for mechanical work in an isothermal process.

55. *The Thomson equation* $\psi = E + t\left(\dfrac{\partial \psi}{\partial t}\right)_v$.

Since $d\psi = -pdv - \phi dt$, $\dfrac{\partial \psi}{\partial t} = -\phi$ when v is constant;

or $\left(\dfrac{\partial \psi}{\partial t}\right)_v = -\phi$.

Now $\qquad\qquad \psi = E - t\phi$.

Therefore $\qquad\qquad \psi = E + t\left(\dfrac{\partial \psi}{\partial t}\right)_v$.

This equation was discovered by Thomson[1] in 1855 and has become fundamental in chemical physics through the wide and varied applications made by Gibbs, Helmholtz, Van 't Hoff, Nernst and others.

The function ψ was reintroduced in 1875 by Willard Gibbs as 'the characteristic function at constant temperature' and by Helmholtz as the '*free energy*'; in 1869, Massieu[2] had used a function (characteristic function), which was in effect $-\psi/t$, to deduce the thermodynamical properties of a fluid.

56. *Thermodynamic potential at constant pressure* (ζ). This is defined by the equation

$$\zeta = E - t\phi + pv,$$

and was used by Gibbs and (effectively) Massieu.

$$d\zeta = dE - td\phi - \phi dt + pdv + vdp$$
$$= vdp - \phi dt.$$

---

[1] *Coll. Papers*, vol. I. p. 297.
[2] *C.R.* LXIX. 1869.

57. *The equilibrium of a liquid in contact with its vapour.*
Let m, v, ϕ, E, t, p be the mass, volume of unit mass, entropy, internal energy of unit mass, temperature, pressure of the liquid and let corresponding dashed letters be the same for the vapour. Let the mixture be contained in a heat-proof envelope of constant volume. It is proposed to find the conditions that the mixture may be in equilibrium.

By the first law, $dQ = dE + dW$, and here dQ and dW are zero, so that E is constant for the mixture. Apply the principle of Clausius that the entropy tends to a maximum. Then since the volume, mass, internal energy are constant, we have to make $m\phi + m'\phi'$ a maximum subject to $mv + m'v' =$ constant, $m + m' =$ constant, and $mE + m'E' =$ constant.

$$\left. \begin{aligned} \therefore \quad md\phi + \phi dm + m'd\phi' + \phi'dm' &= 0 \\ mdv + vdm + m'dv' + v'dm' &= 0 \\ dm + dm' &= 0 \\ mdE + Edm + m'dE' + E'dm' &= 0 \end{aligned} \right\} .$$

Also
$$dE = td\phi - pdv.$$

$$\left. \begin{aligned} \therefore \quad md\phi + \phi dm + \ldots &= 0 \\ mdv + vdm + \ldots &= 0 \\ dm + \ldots &= 0 \\ m\,(td\phi - pdv) + Edm + \ldots &= 0 \end{aligned} \right\} .$$

Using multipliers λ, μ, ν, 1 for the equations, adding, and equating to zero the coefficients of the various differentials, we have

$$\left. \begin{aligned} \lambda + t &= 0 \\ \lambda + t' &= 0 \end{aligned} \right\}, \quad \left. \begin{aligned} \mu - p &= 0 \\ \mu - p' &= 0 \end{aligned} \right\}, \quad \left. \begin{aligned} \lambda\phi + \mu v + \nu + E &= 0 \\ \lambda\phi' + \mu v' + \nu + E' &= 0 \end{aligned} \right\}.$$

Therefore

$$\left. \begin{aligned} t &= t' \\ p &= p' \end{aligned} \right\} \text{ and } E - t\phi + pv = -\nu = E' - t'\phi' + p'v'$$

or
$$\zeta = \zeta'.$$

Thus the liquid and the vapour must be at the same temperature and pressure, and the function ζ must be the same for both.

'It should be noted that the assumption of a rigid, non-conducting envelope enclosing the mixture involves no loss of generality in the result, for if any mass is in equilibrium it would be also if the whole or any part of it were enclosed in such an envelope; therefore the conditions of equilibrium of a mass so enclosed are the general conditions which must always be satisfied in any case of equilibrium' (Gibbs).

58. In the theory of heat-engines, the function $-\zeta$ is usually denoted by G (after Gibbs), so that $G = t\phi - I$. Since G is the same for the liquid and the vapour whatever their proportions, the constancy of G for a wet mixture (steam and water) is a property of the greatest use in calculating I, in terms of which the efficiency of the cycle is usually expressed.

59. *Boundary curves on the ϕ-t diagram.*

The water boundary curve indicates the entropy of water at different temperatures when steam is about to form;

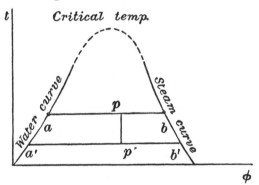

the steam boundary curve indicates the entropy of saturated steam at different temperatures when steam is about to condense.

Consider a line of constant temperature ab. Any point p on it represents a state of mixture of steam and water. Let x be the 'dryness' of the steam, *i.e.* the mass of steam per unit mass of the mixture.

The area under ab is the heat to vaporise unit mass of water and the area under pb is the heat to vaporise $(1 - x)$ of water and these areas are as $ab : pb$.

Therefore $$\frac{1}{1 - x} = \frac{ab}{pb} \text{ or } x = \frac{ap}{ab},$$

so that the point p indicates the 'dryness' of the steam.

60. *Adiabatic expansion of steam.* If the steam is in the state p and expands adiabatically, *i.e.* along pp' at constant entropy to the state p', the dryness $a'p'/a'b'$ at the new temperature can be read off from the diagram.

CHAPTER VI

THE RANKINE CYCLE; REFRIGERATION

61. *The Rankine cycle.*

This is a reversible cycle which represents the working of an ideal engine with the boiler, cylinder, and condenser of an actual engine, without the irreversible losses due to conduction, throttling and so on which occur in an actual engine.

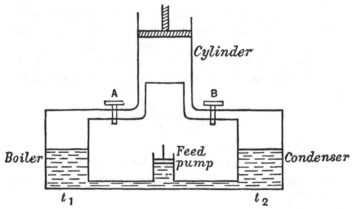

The figure is a diagram of the engine; the p-v and ϕ-t diagrams of the cycle are given on the next page. Suppose the steam which enters the cylinder from the boiler to be dry saturated steam. The cycle is as follows: a represents the state of water returned to the boiler from the condenser by the feed pump. It is at the temperature t_2 of the condenser and at the pressure p_1 of the boiler. It is now heated in the boiler at constant pressure p_1 up to temperature t_1 at which steam begins to form; it is then in the state b. The water is vaporised in the boiler at temperature t_1 and pressure p_1; c is its state as dry saturated steam. The valve A admits the steam to the cylinder where it expands adiabatically to the state d at which it has cooled to the temperature t_2 of the condenser. (It is then a wet mixture whose dryness is estimated by the position of d relative to the two 'boundary' curves within which it lies.)

The valve B opens and admits the wet mixture to the condenser, where it is completely condensed at temperature t_2 and pressure p_2. This is the state e. The feed pump forces the water into the boiler, which it enters at temperature t_2 and pressure p_1, the state a. This completes the cycle.

(It should be observed that ma is very small compared with mc in the p-v diagram. If, for instance, the temperatures of the boiler and condenser are 200° and 40° C., ma is the volume of 1 lb. of water at 40° C. and mc is the volume of 1 lb. of dry saturated steam at 200° C.; the tables give

$$\frac{ma}{mc} = \frac{\cdot 01614}{2 \cdot 0738} = \frac{1}{128} \text{ about.)}$$

62. *Efficiency of the cycle.*

The heat is taken in from the boiler at constant pressure along the path abc and is therefore equal to $I_c - I_a$. [§ 53, (i).]

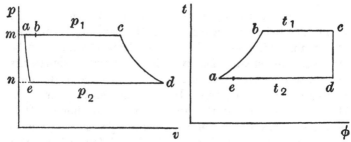

The work done in the cycle is equal to the area of either diagram. Using the p-v diagram, the work done = (area $mcdn$) − (area $maen$). Now the area $mcdn$ is the heat-drop for the adiabatic expansion cd and is therefore equal to

$I_c - I_d$. [§ 53, (ii).] As to the area *maen*, the curve *ae* is an isothermal for water and is nearly parallel to the axis of p, so that the figure is a narrow rectangle. Neglecting the difference between *ma* and *ne* and denoting each by v_{w2}, the volume of 1 lb. of water at temperature t_2 at its saturation pressure p_2, the area *maen* $= (p_1 - p_2) v_{w2}$. (It is assumed that the volume of 1 lb. of water at temperature t_2 is the same for pressure p_1 as for pressure p_2.)

The work done is therefore

$$I_c - I_d - (p_1 - p_2) v_{w2}.$$

Therefore the efficiency is

$$\frac{I_c - I_d - (p_1 - p_2) v_{w2}}{I_c - I_a}.$$

Now along the path *ae*, the temperature is constant and t_2.

But, $\qquad dI = td\phi + vdp.$

$$\therefore I_a - I_e = t_2 (\phi_{wa} - \phi_{we}) + \int_e^a vdp$$

$$= t_2 (\phi_{wa} - \phi_{we}) + (\text{area } maen)$$

$$= t_2 (\phi_{wa} - \phi_{we}) + (p_1 - p_2) v_{w2},$$

where ϕ_{wa} is the entropy of water in the state a, (temperature t_2, pressure p_1), and ϕ_{we} is the entropy of water in the state e, (temperature t_2, pressure p_2).

Neglecting this difference, just as we neglected a corresponding difference for v_w,

$$I_a - I_e = (p_1 - p_2) v_{w2}.$$

Therefore the efficiency of the cycle is

$$\frac{I_c - I_d - (p_1 - p_2) v_{w2}}{I_c - I_e - (p_1 - p_2) v_{w2}}.$$

The term $(p_1 - p_2) v_{w2}$, the 'feed-pump' term, is small compared with $I_c - I_d$ or $I_c - I_e$, and may be neglected

in any but very accurate calculations. The efficiency is
thus very approximately

$$\frac{I_c - I_d}{I_c - I_e}.$$

I_c is the total heat of saturated steam at temperature t_1;
I_e is the total heat of water at saturation pressure at
temperature t_2: these are given in the steam tables.

I_d is not however directly given and is found by the use
of the function G.

63. *Calculation of I_d by the use of G.*

The function $G \equiv t\phi - I$ was shown to be constant for
all states of a steam mixture at constant temperature (and
pressure). (§§ 57, 58.)

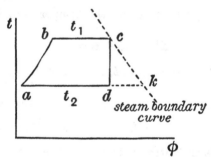

Through c draw the steam boundary curve and let ad
produced meet it in k. Then by the above property of G,

$$G_d = G_k.$$
$$\therefore (t\phi - I)_a = (t\phi - I)_k.$$
$$\therefore t_2\phi_d - I_d = t_2\phi_k - I_k.$$

But
$$\phi_d = \phi_c.$$
$$\therefore I_d = I_k + t_2 (\phi_c - \phi_k).$$

I_k is the total heat of saturated steam at temperature t_2;
ϕ_c is the entropy of saturated steam at temperature t_1; and
ϕ_k is the entropy of saturated steam at temperature t_2.

All these are given in the steam tables, and I_d can be calculated.

64. *Calculation of the efficiency of a Rankine cycle with the steam initially dry and saturated.*

Suppose the steam is initially at 200 lb. pressure and it is condensed at a pressure of 1 lb. (per sq. inch). The steam tables give the corresponding temperatures as 194°·35 C. and 38°·74 C.

Hence

$$t_1 = 273·1 + 194·35 = 467·45, \quad t_2 = 311·74 \text{ abs.}$$

v_{w2} = volume of 1 lb. of water at saturation pressure at 38°·74 = ·01613 c. ft.

$$\therefore (p_1 - p_2) v_{w2} = 199 \times 144 \times ·01613 \text{ ft.-lb.}$$

$$= \frac{199 \times 144 \times ·01613}{1400} \text{ calories} = ·33 \text{ calorie.}$$

Also I_c = total heat of 1 lb. of saturated steam at 194°·35
$$= 669·69 \text{ calories.}$$

I_k = total heat of 1 lb. of saturated steam at 38°·74
$$= 612·46 \text{ calories.}$$

ϕ_c = entropy of 1 lb. of saturated steam at 194°·35
$$= 1·5538 \text{ calories/deg. C.}$$

ϕ_k = entropy of 1 lb. of saturated steam at 38°·74
$$= 2·0067 \text{ calories/deg. C.}$$

I_e = total heat of 1 lb. of water at saturation pressure
$$\text{at } 38°·74 = 37·76 \text{ calories.}$$

Now
$$I_d = I_k + t_2 (\phi_c - \phi_k)$$
$$= 612·46 + 311·74 (- ·4529)$$
$$= 612·46 - 141·0$$
$$= 471·46.$$

$$\therefore I_c - I_d = 198·23, \text{ and } I_c - I_e = 631·93.$$

Therefore the efficiency $= \dfrac{198·23 - ·33}{631·93 - ·33}$

$$= \dfrac{197·9}{631·6}$$

$$= ·313.$$

The efficiency of a Carnot cycle for these temperatures would be

$$1 - \frac{t_2}{t_1} = 1 - \frac{311 \cdot 74}{467 \cdot 45} = 1 - \cdot 665 = \cdot 335.$$

The Rankine cycle, though reversible, is less efficient than the Carnot cycle because some of the heat is not taken in at the *highest* level of temperature t_1, as in Carnot's cycle. Along ab when heat is being taken in the temperature is rising from t_2 to t_1.

65. *The cycle with wet steam or with superheated steam.* If the steam is wet or is superheated on admission to the

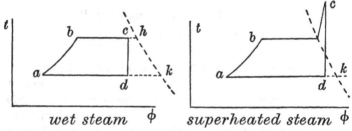

wet steam $\quad\phi\quad$ superheated steam $\quad\phi$

cylinder, the Rankine ϕ-t diagrams take the forms shown in the figures. The dotted line is the steam boundary curve.

Neglecting the feed-pump term, the efficiency is

$$\frac{I_c - I_d}{I_c - I_a}$$

in each case.

In the former case, if q is the dryness of the steam and L_1 is the latent heat at temperature t_1

$I_h - I_c =$ heat taken in as the mixture passes from c to h
$= (1 - q) L_1$.

Thus I_c is found, from the tabulated value of I_h.

In the second case I_c is known from the tables for superheated steam.

In both cases I_d is found by the use of the property of the function G that $G_d = G_k$, in the same way as before.

66. *Mollier I-ϕ diagram.* This diagram, in which I and ϕ are the ordinate and abscissa, is of the utmost value in practical work. In the wet region between the boundary curves, the lines of constant pressure and constant dryness are drawn, and beyond, in the region of superheat, the lines of constant pressure and of constant temperature are shown.

Since $dI = td\phi + vdp$, $\dfrac{\partial I}{\partial \phi} = t$, when p is constant.

Therefore, in the wet region, along a line of constant pressure (which is also a line of constant temperature) $dI/d\phi$ is constant; thus the lines of constant pressure are straight and their gradient is equal to the temperature.

On such a diagram, the points a, b, c, d of a Rankine cycle, for instance, can be readily plotted; the values of I at these points are read off and the efficiency calculated. The dryness of the steam at any stage is also indicated.

Another form of diagram in which I, p are the co-ordinates was also introduced by Mollier and is given with Callendar's steam tables (1915).

67. *Refrigeration.* If a Carnot engine is reversed, it takes heat Q_2 from a cold body at temperature t_2 and gives out Q_1 to a hot one at temperature t_1, and the excess of Q_1 over Q_2 is the work W applied to the piston from outside to drive the engine. It is thus acting as a refrigerator, withdrawing heat from a cold substance.

The 'coefficient of performance' of the refrigerator is the ratio of the cooling effect to the work done, and is therefore

$$\frac{Q_2}{W} \quad \text{or} \quad \frac{Q_2}{Q_1 - Q_2} \quad \text{or} \quad \frac{t_2}{t_1 - t_2}, \quad \text{since} \quad \frac{Q_1}{t_1} = \frac{Q_2}{t_2}.$$

The best effect is obtained when $t_1 - t_2$ is as small as possible.

Vapour compression machine.

The working substance is usually ammonia or carbonic acid. The piston rises, the valve B opens and a cold wet mixture of the working substance and its vapour is taken in from R. The piston falls, B closes, and when the adiabatic compression has raised the pressure sufficiently, the

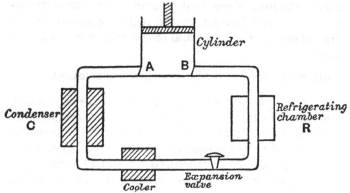

(In the actual machine the tubes in C, R are coils of piping.)

valve A opens and the substance, now at a higher temperature, is discharged into C, kept cool by running water. It is still further cooled by a 'cooler' to a temperature not much above that in R. It then passes the 'expansion valve' when its pressure suddenly falls to that of R. It then evaporates at this low pressure and cools the refrigerating chamber by abstracting heat from it. The cycle is then complete.

68. The ϕ-t diagram is

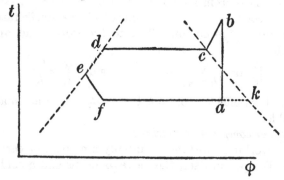

The dotted curves *de*, *ck* are the liquid and vapour boundary curves.

a is the state of the substance drawn into the cylinder from *R* as the piston rises,

ab is the compression in the cylinder,

bcd the condensation in the condenser to liquid at *d*,

de is the cooling of the liquid in the cooler, and

ef is a curve of constant 'total heat' representing the throttling process at the expansion valve.

fa is the expansion in the refrigerator.

The heat taken in by the working substance is $I_a - I_f$, and the heat given out is $I_b - I_e$, for the stage *fa* and the stage *bcde* are each conducted at constant pressure. [§ 53 (i).]

But I is constant along *ef*. [§ 53 (iii).]

$$\therefore I_f = I_e.$$

Therefore the heat taken in $= I_a - I_e$,

and the heat given out $= I_b - I_e$.

Therefore the work done W is the difference, which is $I_b - I_a$, the heat drop in the cylinder, as it should be.

The coefficient of performance is therefore $\dfrac{I_a - I_e}{I_b - I_a}$.

I_a is connected with I_k in the usual manner by the use of the function G and the result is then calculated from the tables.

Or, the cycle can be marked on a Mollier diagram and the values of the various I's read off.

CHAPTER VII

THERMODYNAMICAL RELATIONS FOR A SIMPLE SUBSTANCE

69. The quantities p, v, t, ϕ, E, ... are functions of the state of the substance and any two may be taken as independent variables.

If p, v are used as variables, $t = f(p, v)$,

$$dt = \frac{\partial t}{\partial p}\, dp + \frac{\partial t}{\partial v}\, dv;$$

if p, ϕ are used as variables, $t = F(p, \phi)$,

$$dt = \frac{\partial t}{\partial p}\, dp + \frac{\partial t}{\partial \phi}\, d\phi.$$

The values of $\dfrac{\partial t}{\partial p}$ in the two equations are different and a notation is necessary to distinguish them; the former is denoted by $\left(\dfrac{\partial t}{\partial p}\right)_v$, being the value of $\dfrac{\partial t}{\partial p}$ when v is the other independent variable; the latter by $\left(\dfrac{\partial t}{\partial p}\right)_\phi$, being the value of $\dfrac{\partial t}{\partial p}$ when ϕ is the other independent variable.

70. *Theorems on partial differentiation.*

(a) Let it be given that $dx = Y\,dy + Z\,dz$, where x is a function of y and z.

Then $\qquad \left(\dfrac{\partial Y}{\partial z}\right)_y = \left(\dfrac{\partial Z}{\partial y}\right)_z.$

To prove this, we have

$$x = f(y, z).$$

$$\therefore\ dx = \left(\frac{\partial f}{\partial y}\right)_z dy + \left(\frac{\partial f}{\partial z}\right)_y dz.$$

$$\therefore\ Y = \left(\frac{\partial f}{\partial y}\right)_z \text{ and } Z = \left(\frac{\partial f}{\partial z}\right)_y.$$

$$\therefore\ \left(\frac{\partial Y}{\partial z}\right)_y = \frac{\partial^2 f}{\partial y\,\partial z} = \left(\frac{\partial Z}{\partial y}\right)_z.$$

(β) If x, y, z are connected by a functional equation so that each is a function of the other two, then

(i) $\left(\frac{\partial z}{\partial x}\right)_y = 1 \Big/ \left(\frac{\partial x}{\partial z}\right)_y,$

and similar relations;

(ii) $\left(\frac{\partial y}{\partial z}\right)_x \left(\frac{\partial z}{\partial x}\right)_y \left(\frac{\partial x}{\partial y}\right)_z = -1.$

To prove these, we have

$$dx = \left(\frac{\partial x}{\partial y}\right)_z dy + \left(\frac{\partial x}{\partial z}\right)_y dz$$

and

$$dz = \left(\frac{\partial z}{\partial x}\right)_y dx + \left(\frac{\partial z}{\partial y}\right)_x dy$$

Therefore eliminating dz,

$$dx = \left(\frac{\partial x}{\partial y}\right)_z dy + \left(\frac{\partial x}{\partial z}\right)_y \left[\left(\frac{\partial z}{\partial x}\right)_y dx + \left(\frac{\partial z}{\partial y}\right)_x dy\right].$$

Now dx, dy are independent.

Therefore $\quad 1 = \left(\frac{\partial x}{\partial z}\right)_y \left(\frac{\partial z}{\partial x}\right)_y,$

which is relation (i), and

$$0 = \left(\frac{\partial x}{\partial y}\right)_z + \left(\frac{\partial x}{\partial z}\right)_y \left(\frac{\partial z}{\partial y}\right)_x,$$

or using (i), $\quad \left(\frac{\partial y}{\partial z}\right)_x \left(\frac{\partial z}{\partial x}\right)_y \left(\frac{\partial x}{\partial y}\right)_z + 1 = 0,$

which is relation (ii).

71. Maxwell's equations. We have seen that

$$dE = t\,d\phi - p\,dv,$$
$$dI = t\,d\phi + v\,dp, \qquad \text{(§§ 52–56)}$$
$$d\psi = -p\,dv - \phi\,dt,$$
$$d\zeta = v\,dp - \phi\,dt.$$

Using theorem (a) § 70, it follows that

$$\left.\begin{array}{r}\left(\dfrac{\partial t}{\partial v}\right)_\phi = -\left(\dfrac{\partial p}{\partial \phi}\right)_v \\[2mm] \left(\dfrac{\partial t}{\partial p}\right)_\phi = \left(\dfrac{\partial v}{\partial \phi}\right)_p \\[2mm] \left(\dfrac{\partial p}{\partial t}\right)_v = \left(\dfrac{\partial \phi}{\partial v}\right)_t \\[2mm] \left(\dfrac{\partial v}{\partial t}\right)_p = -\left(\dfrac{\partial \phi}{\partial p}\right)_t \end{array}\right\} \quad \text{(Maxwell).}$$

72. Specific heat equations.

$dQ = dE + pdv$, and E is a function of two of p, v, t.

Hence $\qquad dE = \dfrac{\partial E}{\partial t}\, dt + \dfrac{\partial E}{\partial v}\, dv.$

Therefore dQ is of the form $Pdt + ldv$, and since

$$\dfrac{\partial Q}{\partial t} = c_v, \text{ (the specific heat at constant volume)},$$

$$P = c_v$$

we may write

$$dQ = c_v dt + ldv.$$

dQ is not the differential of a function Q of the variables, but it is equal to $td\phi$, where $d\phi$ is the differential of the entropy ϕ.

$$\therefore \ td\phi = c_v dt + ldv.$$

$$\therefore \ l = t\left(\dfrac{\partial \phi}{\partial v}\right)_t$$

$$= t\left(\dfrac{\partial p}{\partial t}\right)_v, \text{ by Maxwell's third equation.}$$

$$\therefore \ d\phi = \dfrac{c_v}{t}\, dt + \left(\dfrac{\partial p}{\partial t}\right)_v dv.$$

Using theorem (a) § 70, we have

$$\left[\dfrac{\partial}{\partial v}\left(\dfrac{c_v}{t}\right)\right]_t = \dfrac{\partial}{\partial t}\left[\left(\dfrac{\partial p}{\partial t}\right)_v\right]_v,$$

or $\qquad \dfrac{1}{t}\left(\dfrac{\partial c_v}{\partial v}\right)_t = \left(\dfrac{\partial^2 p}{\partial t^2}\right)_v.$

In a similar manner,
$$dQ = c_p \, dt + l' dp,$$
whence using Maxwell's fourth equation
$$l' = -t \left(\frac{\partial v}{\partial t} \right)_p.$$

The use of theorem (a) then leads to
$$\frac{1}{t} \left(\frac{\partial c_p}{\partial p} \right)_t = - \left(\frac{\partial^2 v}{\partial t^2} \right)_p.$$

73. *Variation of c_v with volume at constant temperature.* Kuenen has shown experimentally that c_v increases with v to a maximum and then decreases.

Now
$$\left(\frac{\partial c_v}{\partial v} \right)_t = t \left(\frac{\partial^2 p}{\partial t^2} \right)_v.$$

From Van der Waals' equation, $\left(\frac{\partial^2 p}{\partial t^2} \right)_v = 0$ and therefore $\left(\frac{\partial c_v}{\partial v} \right)_t = 0$, so that $c_v = $ constant at constant temperature, so that the equation fails in this particular.

The maximum of c_v corresponds to
$$\left(\frac{\partial c_v}{\partial v} \right)_t = 0 \ \text{ or } \ \left(\frac{\partial^2 p}{\partial t^2} \right)_v = 0.$$

Dieterici's equation gives a variation of c_v with v, but $\left(\frac{\partial^2 p}{\partial t^2} \right)_v$ cannot vanish, so that it fails to disclose the maximum. Lees' equation, however, does give a maximum in agreement with Kuenen for a reduced temperature near the value 2.

74. *Further properties of the specific heats.*
(i) Since $c_v dt + l dv = dQ = c_p dt + l' dp$
$$(c_p - c_v) \, dt = l dv - l' dp.$$
$$\therefore \ c_p - c_v = l \frac{\partial v}{\partial t} \text{ when } dp = 0$$
$$= l \left(\frac{\partial v}{\partial t} \right)_p.$$
$$\therefore \ c_p - c_v = t \left(\frac{\partial p}{\partial t} \right)_v \left(\frac{\partial v}{\partial t} \right)_p.$$

(ii) Now by theorem (β)

$$\left(\frac{\partial v}{\partial p}\right)_t \left(\frac{\partial p}{\partial t}\right)_v \left(\frac{\partial t}{\partial v}\right)_p = -1.$$

$$\therefore \left(\frac{\partial v}{\partial p}\right)_t \left(\frac{\partial p}{\partial t}\right)_v = -\left(\frac{\partial v}{\partial t}\right)_p.$$

$$\therefore c_p - c_v = -t \left(\frac{\partial v}{\partial t}\right)_p^2 \bigg/ \left(\frac{\partial v}{\partial p}\right)_t.$$

Since increase of pressure leads to decrease of volume, at constant temperature, $\left(\frac{\partial v}{\partial p}\right)_t$ is negative.

$$\therefore c_p - c_v \text{ is positive, or } c_p > c_v.$$

It also appears that $c_p = c_v$ when $\left(\frac{\partial v}{\partial t}\right)_p = 0$, i.e. at the temperature of maximum density, e.g. 4° C. for water.

(iii) Again

$$t d\phi = c_v dt + l dv = c_p dt + l' dp.$$

$$\therefore d\phi = \frac{c_v}{t} dt + \left(\frac{\partial p}{\partial t}\right)_v dv = \frac{c_p}{t} dt - \left(\frac{\partial v}{\partial t}\right)_p dp,$$

using the values found for l, l'.

From these

$$\frac{\partial t}{\partial v} = -\frac{t}{c_v}\left(\frac{\partial p}{\partial t}\right)_v \Bigg\}$$

and

$$\frac{\partial t}{\partial p} = \frac{t}{c_p}\left(\frac{\partial v}{\partial t}\right)_p \Bigg\} \text{ when } \phi \text{ is constant.}$$

$$\therefore \left(\frac{\partial t}{\partial v}\right)_\phi = -\frac{t}{c_v}\left(\frac{\partial p}{\partial t}\right)_v$$

$$\left(\frac{\partial t}{\partial p}\right)_\phi = \frac{t}{c_p}\left(\frac{\partial v}{\partial t}\right)_p.$$

Therefore if $\gamma = \frac{c_p}{c_v}$,

$$\gamma = -\frac{\left(\frac{\partial v}{\partial t}\right)_p \left(\frac{\partial t}{\partial v}\right)_\phi}{\left(\frac{\partial t}{\partial p}\right)_\phi \left(\frac{\partial p}{\partial t}\right)_v}$$

$$= -\left(\frac{\partial p}{\partial v}\right)_\phi \left(\frac{\partial v}{\partial t}\right)_p \left(\frac{\partial t}{\partial p}\right)_v.$$

and $\left(\dfrac{\partial v}{\partial t}\right)_p \left(\dfrac{\partial t}{\partial p}\right)_v \left(\dfrac{\partial p}{\partial v}\right)_t = -1.$

$$\therefore\ \gamma = \dfrac{\left(\dfrac{\partial p}{\partial v}\right)_\phi}{\left(\dfrac{\partial p}{\partial v}\right)_t} = \dfrac{\text{gradient of an adiabatic}}{\text{gradient of an isothermal}}$$

on the p-v diagram.

Since $\gamma > 1$, the adiabatics are steeper than the isothermals, in general.

75. *The equation* $c_p - c_v = \dfrac{vt\alpha^2}{\kappa}.$

The 'coefficient of expansion' α of the substance at constant pressure is the increase of unit volume per unit increase of temperature, which is

$$\frac{1}{v}\left(\frac{\partial v}{\partial t}\right)_p.$$

The 'compressibility' κ of the substance at constant temperature is the decrease of unit volume per unit increase of pressure, which is

$$-\frac{1}{v}\left(\frac{\partial v}{\partial p}\right)_t.$$

Hence from § 74 (ii) $c_p - c_v = \dfrac{vt\alpha^2}{\kappa}.$

While c_p is readily found by experiment and c_v is only found with great difficulty directly, this formula is of value in determining c_v from observed values of c_p. Rankine first used this formula for this purpose as also Nernst and Lindemann[1] (1911), who for the purpose of their work on the variation of c_v with *temperature*, determined the values of c_v from those of c_p for a wide range of temperatures.

76. As an illustration, take the case of mercury at 0° C. By experiment, it is found that

$\alpha = \cdot 000179$ per degree C.

$\kappa = \cdot 00000338$ per atmosphere.

$c_p = \cdot 0333$ heat units.

---

[1] *Zeits. für Elektrochem.* 1911.

Also $v = \dfrac{1}{\rho} = \dfrac{1}{13\cdot596}$ ·

$t = 273\cdot1.$

$\kappa = \dfrac{\cdot00000338}{1013600}$ per dyne per sq. cm.

$\therefore \; c_p - c_v = \dfrac{vta^2}{\kappa}$ ergs/deg.

$$= \dfrac{\dfrac{273\cdot1}{13\cdot596}(\cdot000179)^2 \dfrac{1013600}{\cdot00000338}}{41\cdot8 \times 10^6} \text{ calories/deg.}$$

$= \cdot0046.$

$\therefore \; c_v = \cdot0333 - \cdot0046$

$\qquad = \cdot0287.$

Thus $\dfrac{c_p}{c_v} = 1\cdot16$ for mercury at 0° C., or $\gamma = 1\cdot16$.

γ is greater for gases than for solids and liquids; for the *vapour* of mercury $\gamma = 1\cdot33$.

The formulae of this chapter have been tested by the experiments of Bridgman[1] who for water, and for many other liquids, obtained the volume of one gram at temperatures and pressures ranging from 0° to 80° C. and zero to 12500 kilograms per sq. cm.

77. *Calculation of E, ϕ, ... for a simple system.*

$dQ = dE + pdv$ and $dQ = c_v dt + ldv$, where $l = t\left(\dfrac{\partial p}{\partial t}\right)_v$.

$\therefore \; dE = c_v dt + (l - p)\,dv$

$\qquad = c_v dt + \left\{t\left(\dfrac{\partial p}{\partial t}\right)_v - p\right\} dv$

$\qquad = f(v, t)\,dt + F(v, t)\,dv,$

where $\qquad f(v, t) = c_v,$

and $\; F(v, t) = t\left(\dfrac{\partial p}{\partial t}\right)_v - p.$

[1] *Proc. Amer. Acad. Sc.* 1912, 1913.

Let E be E_1 for the state A (v_1, t_1)

and E_2 for the state B (v_2, t_2).

Then $E_2 - E_1 = \int_A^B f(v, t)\, dt + \int_A^B F(v, t)\, dv.$

The value is the same for all paths from A to B.

Choose the path ALB, so that along AL, $dv = 0$ and along LB, $dt = 0$.

Then $E_2 - E_1 = \int_A^L f(v_1, t)\, dt + \int_L^B F(v, t_2)\, dv$

$$= \int_{t_1}^{t_2} f(v_1, t)\, dt + \int_{v_1}^{v_2} F(v, t_2)\, dv.$$

These are simple integrals of known functions, one with respect to t and the other with respect to v. If the state 1 is a standard state and 2 any state, E is determined for all states.

CHAPTER VIII

THE JOULE-THOMSON POROUS PLUG
EXPERIMENT[1]

78. Mayer in 1842 found a value for the mechanical
equivalent of heat based on the assumption that when a
gas is compressed, the heat developed is equal to the work
done in the compression. To test this hypothesis, Thomson
proposed the porous plug experiment, by which he and
Joule carried out a long series of determinations from 1852
to 1862. Their results proved that when air is compressed
at 10° C. from 1 to 4·7 atmospheres, the heat evolved
exceeds the work spent by $\frac{1}{174}$. For carbonic acid the
fraction was $\frac{1}{32}$, while for hydrogen the heat evolved was
less than the work spent by $\frac{1}{630}$.

Thus it appears that when air is compressed, part of
the heat evolved comes from the store of internal energy
of the gas.

The experiment consisted in pumping the gas to be
examined through a pipe, impervious to heat, plugged with
cotton-wool or filaments of silk, the temperature and
pressure of the gas on each side of the plug being accurately
noted. On account of the resistance of the plug the pressure
of the gas was considerably higher before the passage
through the plug than after, so that there was a rapid
expansion during the passage. The gas was found to be
slightly cooled by its passage through the plug, except in
the case of hydrogen, where a slight heating effect was
observed.

The final form of the apparatus used by Joule and
Thomson[2] is shown in the figure appended.

[1] *Phil. Mag.* 1852.
[2] *Phil. Trans.* 1862.

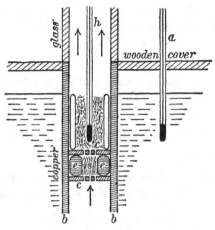

The gas is pumped through a coil of copper pipe immersed in a bath of water whose temperature is measured by the thermometer a. bb is the upper end of the copper pipe. cc are two perforated metallic plates, between which is an indiarubber ring ee containing a silk plug s, which can be compressed to any desired degree by pressing cc together. ff is a cork ring to prevent conduction from the bath, and cotton-wool is loosely packed inside the cork ring. The thermometer h measures the temperature when the gas issues from the plug. The pressure of the gas on entry is measured by a gauge attached to the copper coil and on emergence the pressure is that of the atmosphere.

79. *Theory of the experiment.* The process is a throttling process, so that I is unchanged by the passage of the gas through the plug.

Since
$$dI = d\,(E + pv) = dE + pdv + vdp = dQ + vdp,$$
and $\quad dQ = c_p\, dt + l'dp \quad$ where $l' = -\,t\left(\dfrac{\partial v}{\partial t}\right)_p,$

$$\therefore\ dI = c_p\, dt + \left\{v - t\left(\dfrac{\partial v}{\partial t}\right)_p\right\}dp.$$

$$\therefore\ c_p\left(\dfrac{\partial t}{\partial p}\right)_I = t\left(\dfrac{\partial v}{\partial t}\right)_p - v.$$

In this experiment, I is constant, so that $\left(\dfrac{\partial t}{\partial p}\right)_I$ is the fall of temperature per unit fall of pressure under the conditions of the experiment. This is the measure of the 'cooling effect,' which is therefore equal to

$$\frac{1}{c_p}\left[t\left(\frac{\partial v}{\partial t}\right)_p - v\right].$$

For a perfect gas, $pv = Rt$, and the cooling effect is

$$\frac{1}{c_p}\left[\frac{tR}{p} - v\right],$$

which is zero. Thus for a perfect gas, Mayer's assumption would be exact.

80. *Calculation of the cooling effect for an actual gas.* We use the reduced form of Van der Waals' equation

$$\left(p + \frac{3}{v^2}\right)\left(v - \frac{1}{3}\right) = \frac{8t}{3}.$$

The cooling effect

$$C = \frac{1}{\kappa_p}\left\{t\left(\frac{\partial v}{\partial t}\right)_p - v\right\},$$

where κ_p is the value of c_p in reduced units (κ_p would be the heat required to raise unit mass through a range equal to the critical temperature, the heat being measured in work units such that the unit of work is the product of the critical pressure and the critical volume).

Now $$\left(p + \frac{3}{v^2}\right) - \frac{6}{v^3}\left(v - \frac{1}{3}\right) = \frac{8}{3}\left(\frac{\partial t}{\partial v}\right)_p.$$

$$\therefore \kappa_p . C = \frac{t}{\left(\frac{\partial t}{\partial v}\right)_p} - v$$

$$= \frac{\left(p + \frac{3}{v^2}\right)\left(v - \frac{1}{3}\right)}{\left(p + \frac{3}{v^2}\right) - \frac{6}{v^3}\left(v - \frac{1}{3}\right)} - v$$

$$= \frac{-\frac{p}{3} + \frac{6}{v} - \frac{3}{v^2}}{p - \frac{3}{v^2} + \frac{2}{v^3}}.$$

The denominator is positive, because it is in fact

$$\frac{8}{3}\left(\frac{\partial t}{\partial v}\right)_p$$

which is positive for all gases, since the volume increases with the temperature at constant pressure.

Therefore C is positive if $\dfrac{p}{3} < \dfrac{6}{v} - \dfrac{3}{v^2}$

or $$p < \frac{9\,(2v-1)}{v^2}.$$

81. *Inversion of the cooling effect.* If $p > \dfrac{9\,(2v-1)}{v^2}$, the gas is heated on passing the plug. Thus the 'inversion' of the effect occurs when $p = \dfrac{9\,(2v-1)}{v^2}$; any state of the gas for which p, v satisfy this relation undergoes no change of temperature on passing the plug.

Inversion curve. The curve indicating such states of the gas is called an 'inversion curve'; its form is the same for all gases in the reduced units.

Using the Amagat coordinates $y = pv$, $x = p$, the equation of the inversion curve is

$$x\frac{y^2}{x^2} = 9\left(2\frac{y}{x} - 1\right)$$

or $$y^2 = 9\,(2y - x), \text{ a parabola.}$$

The figure shows the inversion curve, from which it appears that inversion cannot occur if $p > 9$.

There is a cooling effect if the state of the gas is such that $p < \dfrac{9\,(2v-1)}{v^2}$ or $y^2 < 9\,(2y - x)$ or $9x < 18y - y^2$, *i.e.* if the point representing the state (the 'state-point') is inside the curve. For state-points outside, there is a heating effect.

The temperatures of inversion are given by

$$\frac{8t}{3} = \left(p + \frac{3}{v^2}\right)\left(v - \frac{1}{3}\right),$$

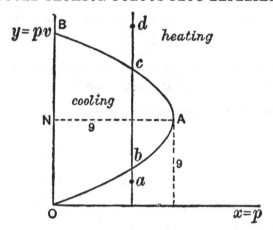

where $$p = 9\left(\frac{2}{v} - \frac{1}{v^2}\right),$$

from which $$\frac{4t}{3} = \left(3 - \frac{1}{v}\right)^2.$$

If P is any point on the inversion curve,

$$v = \frac{y}{x} = \text{the gradient of the line } OP.$$

Now $v > \frac{1}{3}$ (since t must be positive in Van der Waals' equation), or $\frac{1}{v} < 3$.

Therefore t increases as v increases, or as the gradient of OP increases. Hence the temperature increases steadily along the inversion curve as P passes from O to B.

At O, $\frac{1}{v} = 2$ and at B, $\frac{1}{v} = 0$; the corresponding temperatures are therefore $\frac{3}{4}$ and $\frac{27}{4}$. This is the possible range of inversion temperatures for all values of p.

82. By considering a constant pressure line $abcd$ for states between a and d, it appears that for states between a and b there is a heating effect, for those between b and c a cooling effect, and for those between c and d a heating effect. The temperature is increasing from a to d because the volume is increasing at constant pressure. Thus there are two temperatures of inversion for a given pressure; as the temperature increases through the lower of these values the change is from a heating to a cooling effect, and as it increases through the higher of these values the change is from a cooling effect to a heating effect.

Consider the case of hydrogen. The critical temperature is 35° (absolute) and the critical pressure 15 atm. In the experiments of Joule and Thomson the pressure used was 4·7 atm., so that the reduced pressure p was $\frac{47}{150}$. The equation

$$\frac{1}{v^2} - \frac{2}{v} = -\frac{p}{9}$$

gives the corresponding values of $\frac{1}{v}$ as 1·98 and ·02. Using

$$\frac{4}{3}t = \left(3 - \frac{1}{v}\right)^2,$$

we find the values of t as ·78 and 6·66, or 27° abs. and 233° abs. or − 246° C. and − 40° C.

Hence below − 246° C. there would be a heating effect, between − 246° C. and − 40° C. a cooling effect, and above − 40° C. a heating effect. Thus the heating effect observed at ordinary temperatures is accounted for by Van der Waals' equation.

For carbonic acid, whose critical temperature is 304° (absolute) and critical pressure 72 atm., the inversion temperatures at the pressure 4·7 atm. used by Joule and Thomson work out to − 45° C. and 1779° C. Thus at ordinary temperatures there would be a cooling effect.

83. *Deduction of the characteristic equation of a gas from observations of the cooling effect.* Thomson and Joule found[1] that the cooling effect for air and carbonic acid varied as $1/t^2$, within the range of temperatures employed (from

---

[1] Thomson, *Coll. Papers,* vol. I. p. 428.

near 0° C. to near 100° C.); the pressure was much the same in all the experiments, 4 to 5 atmospheres, so that the cooling effect $= a/t^2$, where a is constant for a given pressure, and is in general a function of p.

$$\therefore\ t\left(\frac{\partial v}{\partial t}\right)_p - v = \frac{ca}{t^2}, \text{ where } c \text{ is } c_p.$$

The variations of c with the state of the gas are small in the range of these experiments and these small variations are multiplied by a in the above equation; a being small, their effect is smaller still and they may be neglected. Treating c as constant, we have

$$\left\{\frac{\partial}{\partial t}\left(\frac{v}{t}\right)\right\}_p = \frac{ca}{t^4}.$$

$$\therefore\ \frac{v}{t} = -\frac{ca}{3t^3} + A,$$

where A is a function of p.

$$\therefore\ v = At - \frac{ca}{3t^2}.$$

Now if $a = 0$, the gas is perfect.

$$\therefore\ v = \frac{Rt}{p}. \qquad \therefore\ A = \frac{R}{p}. \qquad \therefore\ v = \frac{Rt}{p} - \frac{ca}{3t^2}.$$

This is the characteristic equation of a gas which would give the cooling effect observed by Thomson and Joule.

84. *Reduction of the readings of an air thermometer to the thermodynamical scale and determination of the thermo-dynamic zero of temperature.* The observations of the cooling effect on air in the porous plug experiment can be used to correct an air thermometer to the absolute Thomson scale.

The equation for the cooling effect is

$$c_p\frac{dt}{dp} = t\left(\frac{\partial v}{\partial t}\right)_p - v;$$

it is based on the absolute Thomson scale.

Let T be the air thermometer temperature corresponding to the Thomson temperature t and let C_p be the specific heat at constant pressure as determined by the use of an air thermometer.

Then $c_p = \dfrac{dQ}{dt}$, $C_p = \dfrac{dQ}{dT}$. $\quad \therefore c_p = C_p \dfrac{dT}{dt}$.

The cooling effect equation becomes

$$C_p \frac{dT}{dt}\left(\frac{dt}{dT}\frac{dT}{dp}\right) = t\left(\frac{\partial v}{\partial T}\right)_p \frac{dT}{dt} - v$$

or

$$C_p \frac{dT}{dp} + v = t\left(\frac{\partial v}{\partial T}\right)_p \frac{dT}{dt}.$$

$$\therefore \int \frac{dt}{t} = \int \left\{\left(\frac{\partial v}{\partial T}\right)_p dT\right\}\Big/\left\{C_p \frac{dT}{dp} + v\right\},$$

$$\therefore \log\left(\frac{t_2}{t_1}\right) = \int_{T_1}^{T_2} \left\{\left(\frac{\partial v}{\partial T}\right)_p dT\right\}\Big/\left\{C_p \frac{dT}{dp} + v\right\}.$$

Now $\dfrac{1}{v}\left(\dfrac{\partial v}{\partial T}\right)_p$ is the coefficient of expansion of air as observed by an air thermometer and $\dfrac{dT}{dp}$ is the Joule-Thomson cooling effect as observed by an air thermometer.

Observation of these for various temperatures enables the integrand on the right-hand side of the equation to be plotted for a given range of temperature T_1 to T_2 on the air thermometer and the integral to be found as an area.

If T_1 and T_2 are the readings of an air thermometer at the temperatures of melting ice and boiling water respectively, the ratio t_2/t_1 for those temperatures is thus found for the absolute scale. Let this ratio thus found be x, and let the interval between the temperatures of melting ice and boiling water be 100 divisions of the absolute scale.

Then $\quad \dfrac{t_2}{t_1} = x$ and $t_2 - t_1 = 100$.

$$\therefore (x-1)t_1 = 100,$$

$$t_1 = \frac{100}{x-1}.$$

Thus the absolute temperature of the melting point of ice is found. The value is 273·1 degrees of the absolute scale. Hence the absolute zero of temperature is 273·1 degrees of the absolute scale below the melting point of ice.

Again, if 0° on the air thermometer is the melting point of ice,

$$\log\left(\frac{t}{273\cdot1}\right) = \int_0^T \left\{\left(\frac{\partial v}{\partial T}\right)_p dT\right\} \Big/ \left\{C_p \frac{dT}{dp} + v\right\}$$

for corresponding temperatures t, T; so that the absolute thermodynamic temperature t corresponding to any temperature T of the air thermometer can be found.

85. *Linde's process for the liquefaction of air.* This process, used on a commercial scale for the liquefaction of air, is based upon the Joule-Thomson cooling effect which occurs when air is allowed to expand through a valve from a high to a low pressure.

Linde arranged for the cooled air to pass out through a pipe enclosing that of the incoming air, so as to pass

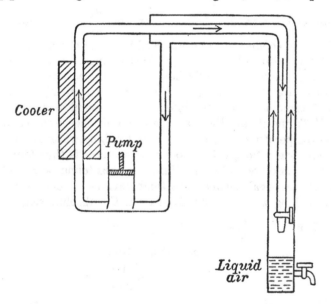

between the two pipes. In this way the cooled air recovers its former temperature and abstracts heat from the incoming air.

The temperature of the incoming air is let down progressively (at compound interest), as the cooled air becomes colder and colder, until liquefaction begins, when the temperature ceases to fall, and the air steadily liquefies.

CHAPTER IX

CLAPEYRON'S EQUATION; CLAUSIUS' EQUATION

86. *Vapour pressure and temperature.* It has been shown (§ 57) that when a liquid is in equilibrium with its vapour, they have a common temperature and pressure, and $\zeta_1 = \zeta_2$. Let the suffix 1 refer to the liquid and the suffix 2 to the vapour and let T, P be their common temperature and pressure.

Then $\qquad E_1 - T\phi_1 + Pv_1 = E_2 - T\phi_2 + Pv_2$.

Now consider an ideal isothermal for the fluid, such as a Van der Waals' isothermal, corresponding to the above temperature T.

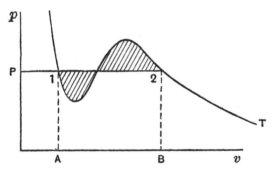

Draw the ordinate $p = P$ meeting this in the points 1, 2. The points 1, 2 represent the states of the liquid and vapour in the above equation.

Let $\qquad\qquad\qquad \psi = E - t\phi$.

Then $\qquad\qquad\qquad d\psi = -\phi dt - pdv$.

Therefore along the isothermal $dt = 0$ and $d\psi = -pdv$.

$$\therefore\ \psi_2 - \psi_1 = -\int_1^2 p\,dv,$$

taken along the isothermal,

$$\therefore\ (E_2 - T\phi_2) - (E_1 - T\phi_1) = -\int_{v_1}^{v_2} p\,dv,$$

along the isothermal.

Using the first equation, we obtain

$$P(v_2 - v_1) = \int_{v_1}^{v_2} p\,dv$$

or the rectangle $12BA$ is equal to the area under the isothermal between 1 and 2, *i.e.* the shaded areas are equal, which is J. Thomson's rule (§ 46).

Using the characteristic equation

$$p = \frac{Rt}{v - b} - \frac{a}{v^2},$$

we have

$$P(v_2 - v_1) = \int_{v_1}^{v_2}\left(\frac{RT}{v - b} - \frac{a}{v^2}\right) dv.$$

$$\left.\begin{aligned}\therefore\ P(v_2 - v_1) &= RT\log\left(\frac{v_2 - b}{v_1 - b}\right) + a\left(\frac{1}{v_2} - \frac{1}{v_1}\right).\\[2mm]\text{Also}\qquad \frac{RT}{v_1 - b} - \frac{a}{v_1^2} &= P = \frac{RT}{v_2 - b} - \frac{a}{v_2^2}.\end{aligned}\right\}$$

These three equations determine P, v_1, v_2 as functions of T. In this way the vapour pressure and densities of the liquid and vapour are found as functions of the temperature.

87. *Clapeyron's equation.* Let p be the vapour pressure at temperature t. The corresponding points 1, 2 satisfy

$$E_1 - t\phi_1 + pv_1 = E_2 - t\phi_2 + pv_2$$

or

$$\zeta_1 = \zeta_2.$$

For an isothermal whose temperature is $t + dt$,

$$\zeta_1 + d\zeta_1 = \zeta_2 + d\zeta_2.$$

$$\therefore\ d\zeta_1 = d\zeta_2.$$

But for a simple substance such as the liquid or the vapour

$$d\zeta = vdp - \phi dt.$$

$$\therefore \quad v_1 dp - \phi_1 dt = v_2 dp - \phi_2 dt.$$

$$\therefore \quad (v_2 - v_1)\frac{dp}{dt} = \phi_2 - \phi_1.$$

Now let L be the latent heat of vaporisation of the liquid at temperature t; then $\dfrac{L}{t}$ = increase of entropy in passing from liquid to vapour at temperature t.

$$\therefore \quad \frac{dp}{dt} = \frac{L}{t(v_2 - v_1)}.$$

This equation, first obtained by Clapeyron[1] from Carnot's theory, has been amply verified by experiment; it has since become of the first importance in the study of the properties of solutions.

88. *Deduction of Clapeyron's equation by the use of a Carnot cycle.* The figure shows two isothermals for a liquid

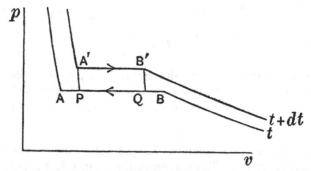

and its vapour for temperatures $t, t + dt$. $A'P$, $B'Q$ are adiabatics through A', B'. Take the substance (mixture of liquid and vapour) round the cycle $A'B'QP$, a Carnot cycle. The heat taken in along $A'B'$ is $L + dL$, the latent heat at temperature $t + dt$.

[1] *Jour. École Polyt.* XIV. 1834.

$\therefore$ the work done in the cycle $A'B'QP$ is

$$(L + dL)\left(1 - \frac{t}{t + dt}\right), \quad \S\, 25,$$

$$= L\,\frac{dt}{t}, \text{ to the first order.}$$

This is equal to the area of the cycle, which

$$= (A'B')\,dp,$$
$$= (AB)\,dp, \text{ ultimately,}$$
$$= (v_2 - v_1)\,dp.$$

$$\therefore \ L\,\frac{dt}{t} = (v_2 - v_1)\,dp.$$

$$\therefore \ \frac{dp}{dt} = \frac{L}{t\,(v_2 - v_1)}.$$

The argument which has led to this result would apply as well to a solid in contact with its own liquid (*e.g.* ice in contact with water); L would be the latent heat of fusion of the solid.

89. *Effect of pressure on the boiling point and the melting point.* For a liquid and vapour

$$\frac{dt}{dp} = \frac{t\,(v_2 - v_1)}{L},$$

where t is the boiling point at pressure p.

Now the volume of unit mass of a liquid (v_1) < the volume of unit mass of its vapour (v_2), so that $\frac{dt}{dp}$ is positive; or the boiling point is raised by increase of pressure.

So, for a solid and liquid,

$$\frac{dt}{dp} = \frac{t\,(v_2 - v_1)}{L},$$

where t is the melting point and v_2, v_1 refer to the liquid and solid.

In the case of ice and water, the volume of unit mass of ice (v_1) > the volume of unit mass of water (v_2).

$$\therefore \; \frac{dt}{dp} \text{ is negative;}$$

or the melting point of ice is lowered by pressure.

90. *Numerical results.*

(i) *To find the latent heat of water boiling at 150° C.*

The pressure needed for water to boil at 150° C. is 69·15 lb. per sq. in. (from tables).

v_2 = volume of 1 lb. of saturated steam at 150° C.

$$= 6\cdot2895 \text{ c. ft.}$$

v_1 = volume of 1 lb. of water at 150° C. = ·0175 c. ft.

Now at 140° C., 150° C., 160° C. the pressures of saturated steam are 52·482, 69·150, 89·800 lb. per sq. in.

The change per degree is 1·667 in the first interval (140°–150°) and is 2·075 in the second (150°–160°).

Taking a mean of these, the change per degree in that region is about 1·87 lb. per sq. in.

Therefore dt/dp, which is the increase of temperature per unit increase in lb. per sq. *foot* of the pressure

$$= \frac{1}{1\cdot87 \times 144}.$$

Also $\qquad\qquad t = 423\cdot1.$

Now $\quad L = t\,(v_2 - v_1)\,\dfrac{dp}{dt}$

$$= 423\cdot1 \times 6\cdot272 \times 1\cdot87 \times 144 \text{ ft.-lb. per lb.}$$

$$= \frac{423\cdot1 \times 6\cdot272 \times 1\cdot87 \times 144}{1400} \text{ calories per lb.}$$

$$= 510\cdot3.$$

The tables give 506·6.

(In the above work, dt/dp was somewhat roughly approximated to from the steam tables; in practice a graph of the variation of t with p is available from which the value of dp/dt can be read off.)

(ii) *To find the lowering of the melting point of ice per atmosphere increase of pressure.* Consider 1 gram of ice.

v_1 = volume of 1 gram of water = 1 c.c.

v_2 = „ „ ice = 1·09 c.c.

L = 79·25 calories per gr. = 79·25 × (41·8 × 10^6) ergs per gr.

$t = 273 \cdot 1.$ $\dfrac{dt}{dp} = \dfrac{-273 \cdot 1 \times \cdot 09}{79 \cdot 25 \times 41 \cdot 8 \times 10^6}.$

If dp = 1 atmosphere = 1013600 dynes per sq. cm.

$$dt = \frac{-273 \cdot 1 \times \cdot 09 \times 1013600}{79 \cdot 25 \times 41 \cdot 8 \times 10^6}$$

$$= - \cdot 00752.$$

This is the lowering of the melting point per atmosphere and agrees with James Thomson's formula, verified experimentally by W. Thomson (p. 3).

91. *Clausius' equation.* Using the figure and notation of §§ 86, 87 we have

$$\frac{L}{t} = \phi_2 - \phi_1 .$$

$$\therefore \frac{d}{dt}\left(\frac{L}{t}\right) = \frac{d\phi_2}{dt} - \frac{d\phi_1}{dt} .$$

$$\therefore \frac{dL}{dt} - \frac{L}{t} = t\frac{d\phi_2}{dt} - t\frac{d\phi_1}{dt} .$$

Now $td\phi_2$ is the heat taken in by the fluid in the state 2 (*i.e.* saturated vapour) when its temperature rises by dt; it is therefore the specific heat of the saturated vapour; denote this by c_2, and denote the specific heat of the liquid at saturation pressure by c_1.

Then $\dfrac{dL}{dt} - \dfrac{L}{t} = c_2 - c_1,$

which is the latent heat equation of Clausius.

92. *Deduction of Clausius' equation by the use of a cycle.*

The curves shown are curves of equal pressure on the ϕ-t diagram for a liquid and its vapour.

The dotted curves are the liquid and vapour boundary curves.

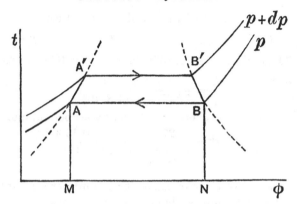

Take the substance round the cycle $A'B'BAA'$.
The heat taken in is

$$(L + dL) - c_2 dt - L + c_1 dt,$$

the four terms corresponding to $A'B'$, $B'B$, BA, AA'.

This is equal to the area of the cycle, which is to $ABNM$ in the ratio $dt : t$.

But $ABNM$ is the heat given out along $BA = L$.

$$\therefore \text{ Area of cycle} = L \frac{dt}{t}.$$

$$\therefore dL - c_2 dt + c_1 dt = L \frac{dt}{t}$$

or
$$\frac{dL}{dt} - \frac{L}{t} = c_2 - c_1.$$

93. *The specific heat of saturated steam.* This is calculated by the use of Clausius' equation

$$\frac{dL}{dt} - \frac{L}{t} = c_2 - c_1.$$

Consider saturated steam at 100° C. and therefore atmospheric pressure.

The steam tables give L at 90° C., 100° C., 110° C. as 545·25, 539·30, 533·17 calories per lb., so that L falls with rising temperature.

The decrease of L per degree is ·595 in the first interval and 613 in the second. Taking the mean as the approximate value, the decrease of L per degree at 100° C. is ·604.

$$\therefore \frac{dL}{dt} = - ·604.$$

Also $\qquad \dfrac{L}{t} = \dfrac{539·30}{373·1} = 1·445$

and $c_1 = 1·013$ for water at 100° C.

Therefore the specific heat of saturated steam at 100° C. (c_2)

$$= c_1 + \frac{dL}{dt} - \frac{L}{t}$$
$$= 1·013 - ·604 - 1·445$$
$$= - 1·036.$$

Thus the specific heat of steam, kept saturated, is negative.

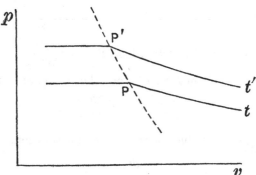

In order to examine this curious result, consider the passage of steam from P to P' along the curve of saturation (dotted) as its temperature rises from t to t'. The steam can be guided along this path by compression and addition of heat. Let the work done on the steam in compression be W and the heat added be Q.

Then if E, E' are the internal energies of the states P, P',

$$E' - E = Q + W \text{ by the first law.}$$

For two given states, $E' - E$ is definite $= \kappa$ say.

$$\therefore \ Q + W = \kappa.$$
$$\therefore \ Q = \kappa - W.$$

If W should happen to be $> \kappa$, then Q is negative and there is a negative specific heat for the change; if $W < \kappa$, there is a positive one.

94. *Expression for the specific heat at constant saturation in terms of the specific heat at constant pressure.*

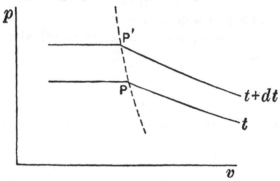

$$dQ = c_p\, dt + l'dp \text{ where } l' = - t \left(\frac{\partial v}{\partial t}\right)_p.$$

For passage from P to P' along the curve of saturation,

$$\frac{dQ}{dt} = c_p + l'\frac{dp}{dt},$$

where $\frac{dp}{dt}$ is the value for the curve of saturation. This is

known to be $\dfrac{L}{t\,(v_2 - v_1)}$ and $\dfrac{dQ}{dt}$ is then c_2.

$$\therefore \; c_2 = c_p - t \left[\left(\frac{\partial v}{\partial t}\right)_p\right]_2 \frac{L}{t\,(v_2 - v_1)},$$

$$c_2 = c_p - L \frac{\left[\left(\frac{\partial v}{\partial t}\right)_p\right]_2}{v_2 - v_1},$$

where $\left[\left(\dfrac{\partial v}{\partial t}\right)_p\right]_2$ is the value of $\left(\dfrac{\partial v}{\partial t}\right)_p$ calculated for un-

saturated steam, with the saturation values v_2, t, p afterwards put in for v, t, p.

CHAPTER X

EQUILIBRIUM OF SYSTEMS;
THE PHASE RULE

95. *Conditions of natural change of a system.*
Isolated system. If a system is isolated from all external influences, any natural (*i.e.* spontaneous) change of state increases the entropy, or $d\phi > 0$.

96. *Isothermal change.* In many of the processes of physical chemistry the system is surrounded by a medium (air, the containing vessel, etc.), the whole being at a common temperature t and the medium so large that the changes of the system have no appreciable effect on the temperature of the medium.

If ϕ, ϕ_0 are the entropies of the system and medium, then in any natural change $d\phi + d\phi_0 > 0$, since the two form an isolated system.

Assuming that any transfer of heat dQ from the medium to the system can be done reversibly, then

$$-\frac{dQ}{t} = d\phi_0,$$

$$\therefore \ d\phi - \frac{dQ}{t} > 0.$$

But for the system
$$dQ = dE + dW.$$

Therefore $\quad td\phi - dE - dW > 0,$

or $\quad\quad\quad td\phi - dE > dW,$

or $\quad\quad\quad d(E - t\phi) < -dW,$

since t is constant, or $\quad d\psi < -dW,$

where dW is the work done by the system during the isothermal change.

(α) *Isothermal change at constant volume.* If the work done during the change is zero, as for example, when the volume of the system remains constant, $d\psi < 0$ is the condition of isothermal change.

(β) *Isothermal change at constant pressure.* If the work done is entirely due to changes of volume, then the condition of isothermal change $d\psi < - dW$ becomes

$$d\psi + pdv < 0.$$

For an isothermal change at constant pressure this becomes

$$d\,(\psi + pv) < 0,$$

or $$d\,(E - t\phi + pv) < 0,$$

or $$d\zeta < 0.$$

Thus in a natural isothermal change at constant volume, $d\psi < 0$ and in a natural isothermal change at constant pressure, $d\zeta < 0$.

It should be observed that in a natural isothermal process (*i.e.* an irreversible one) $d\psi < - dW$ or $dW < - d\psi$, or the work done by the system in passing from a state 1 to a state 2 is $< \psi_1 - \psi_2$.

It has been seen (§ 54) that if the process were reversible, the work done $= \psi_1 - \psi_2$. Thus the decrease of ψ is the maximum work that the system can do in passing from the first state to the second.

97. *Conditions of equilibrium of a system.* If in any virtual change δ of a system the condition for an actual natural change d is not satisfied, the change cannot occur and the system is in stable equilibrium.

Thus if a system at constant temperature and volume is in a state such that any virtual change δ from that state would make $\delta\psi > 0$, it is in stable equilibrium, or ψ must be a minimum for stable equilibrium. So for a system at constant temperature and pressure, ζ must be a minimum for stable equilibrium.

'If the system is isolated from all external influences

and its entropy is greater than in any other state of the same energy (or its energy less than in any other state of the same entropy) it is in stable equilibrium, for any natural change of state must involve either a decrease of entropy or an increase of energy, which are alike impossible for an isolated system' (Gibbs).

98. *Fundamental equation of a simple substance.* Suppose the internal energy E is known as a function of v, ϕ, so that $E = f(v, \phi)$.

Now $\qquad dE = td\phi - pdv.$

Therefore $\qquad t = \dfrac{\partial E}{\partial \phi}, \quad p = -\dfrac{\partial E}{\partial v}.$

Hence $\qquad I = E + pv = E - v\dfrac{\partial E}{\partial v},$

$$\psi = E - t\phi = E - \phi\dfrac{\partial E}{\partial \phi},$$

$$\zeta = E - t\phi + pv = E - \phi\dfrac{\partial E}{\partial \phi} - v\dfrac{\partial E}{\partial v}.$$

Thus t, p, I, ψ, ζ are all deducible in terms of v, ϕ.

Such an equation $E = f(v, \phi)$ is called a 'fundamental equation' (Massieu) for the substance, as from it all the other thermodynamical quantities can be found.

In a similar way, it can be shown that $I = f(p, \phi)$, or $\psi = f(t, v)$ or $\zeta = f(t, p)$ is each a fundamental equation.

99. *Thermodynamic surface for a simple substance.* Willard Gibbs[1] in 1873 showed the great value of the use of geometrical figures in the study of thermodynamical processes. He took the values of v, ϕ, E as the coordinates x, y, z of a point so that the fundamental equation $E = f(v, \phi)$ became $z = f(x, y)$, a surface. The various parts of this surface represent the solid, liquid and vapour states of the substance and this surface Gibbs calls the 'primitive' surface.

---

[1] *Sc. Papers*, vol. I. p. 33.

Since $\qquad t = \dfrac{\partial E}{\partial \phi}, \quad p = -\dfrac{\partial E}{\partial v},$

we have $\qquad t = \dfrac{\partial z}{\partial y}, \quad p = -\dfrac{\partial z}{\partial x},$

so that the temperature and pressure are determined by the inclination of the tangent plane to the planes

$$x = 0, \quad y = 0.$$

[The tangent plane at $x_0 y_0 z_0$ is

$$(x - x_0)\left(\frac{\partial z}{\partial x}\right)_0 + (y - y_0)\left(\frac{\partial z}{\partial y}\right)_0 - (z - z_0) = 0$$

or $\qquad (x - x_0)(-p_0) + (y - y_0)t_0 - z - z_0 = 0.$

It is parallel to the plane $z = -xp_0 + yt_0.$]

From the 'primitive' surface representing the substance in the homogeneous state as solid, liquid or vapour, Gibbs obtained a 'derived surface' consisting of three developable surfaces representing states of mixture of two of the three states solid, liquid, vapour, and a triangular plane representing states of mixture of all three (see Gibbs' *Scientific Papers*, vol. I. pp. 35–38 and pp. 43–45).

These two surfaces determine the thermodynamical properties of the substance in all possible states (§ 100).

100. *Geometrical condition of stability.* Let the substance be in a state of equilibrium represented by P_0 and let P_1 be another state. Let a line through P_1 parallel to Oz meet the tangent plane to the thermodynamic surface at P_0 in N.

The equation of the tangent plane is

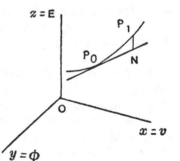

$$(x - x_0)\left(\frac{\partial z}{\partial x}\right)_0 + (y - y_0)\left(\frac{\partial z}{\partial y}\right)_0 - (z - z_0) = 0$$

or $\qquad (x - v_0)(-p_0) + (y - \phi_0)t_0 - (z - E_0) = 0.$

If x_1, y_1, z_1 are the coordinates of P_1, then the z of N is found by writing $x = x_1$, $y = y_1$ in the equation of the tangent plane.

Therefore for N,

$$z = E_0 + (x_1 - v_0)(-p_0) + (y_1 - \phi_0) t_0$$
$$= E_0 + (v_1 - v_0)(-p_0) + (\phi_1 - \phi_0) t_0,$$

and for P_1 $\qquad z = E_1$.

Therefore

$$P_1 N = E_1 - E_0 + p_0 (v_1 - v_0) - t_0 (\phi_1 - \phi_0)$$
$$= (E_1 - t_0 \phi_1 + p_0 v_1) - (E_0 - t_0 \phi_0 + p_0 v_0)$$
$$= \zeta_1 - \zeta_0,$$

for constant temperature and pressure, p_0, t_0.

But the condition of stable equilibrium at constant temperature and pressure is that ζ is a minimum. Hence if P_0 is a state of stable equilibrium, $\zeta - \zeta_0$ must be > 0 for any state P_1 near to P_0 or $P_1 N > 0$; that is, the thermodynamic surface must be convex to the tangent plane at P_0 and above it.

Thus to find states of equilibrium under conditions of constant temperature t_0 and pressure p_0, draw all tangent planes to the surface (primitive and derived) whose inclination corresponds to t_0, p_0, *i.e.* parallel to

$$z = -xp_0 + yt_0.$$

The points of contact will be states of equilibrium and the stability will depend upon the relation of the surface to the tangent plane at each point.

The other three forms of fundamental equation could likewise be used to give thermodynamical surfaces; Gibbs made especial use of the ζ-t-p surface (see p. 115 of his *Scientific Papers*, vol. I).

101. *Homogeneous mixtures.* Such a mixture would be, for example, gaseous hydriodic acid in a state of dissociation into hydrogen and iodine vapour; an electrolyte dissociating in a solvent which ionises it; or a mixture of acetic acid and ethyl alcohol producing some ethyl acetate and water. Each mixture contains different substances but is physically and chemically homogeneous throughout.

The internal energy E of the mixture is a function of ϕ, v and of the masses m_1, m_2, ..., m_n of the n different substances which compose it. Consider unit mass of the mixture, so that $m_1 + m_2 + ... + m_n = 1$.

$$dE = \frac{\partial E}{\partial \phi}\, d\phi + \frac{\partial E}{\partial v}\, dv + \frac{\partial E}{\partial m_1}\, dm_1 + \frac{\partial E}{\partial m_2}\, dm_2 + ...$$

If m_1, m_2, ... remain constant in a change of state, the mixture behaves as a simple substance of constant composition, for which $dE = td\phi - pdv$.

But when dm_1, dm_2, ... vanish,

$$dE = \frac{\partial E}{\partial \phi}\, d\phi + \frac{\partial E}{\partial v}\, dv. \quad \therefore t = \frac{\partial E}{\partial \phi}, \ -p = \frac{\partial E}{\partial v}.$$

Therefore for the mixture,

$$dE = td\phi - pdv + \mu_1 dm_1 + \mu_2 dm_2 + ... \quad ...(1),$$

where μ_1 stands for $\dfrac{\partial E}{\partial m_1}$, and so on.

Thus μ_1 means the increase of energy per unit addition of the substance 1 to the system and was called by Gibbs the 'potential' of that substance.

102. *Fundamental equation for the mixture.* Let
$$E = f(v, \phi, m_1, m_2, ... m_n)$$
where f is a known function.
$$dE = td\phi - pdv + \mu_1 dm_1 + ...$$
$$\therefore t = \frac{\partial E}{\partial \phi}, \quad p = -\frac{\partial E}{\partial v}, \quad \mu_1 = \frac{\partial E}{\partial m_1}, ...$$

Thus the quantities t, p, μ_1, ... upon which it will be seen that the thermal and physical properties of the mixture depend, are found.

103. *Other formulae.* Since the mixture is homogeneous, it follows that if m_1, m_2, ... are multiplied in any ratio,

E, ϕ, v are also, so that E is a homogeneous function of the first degree in ϕ, v, m_1, ... m_n.

Therefore by Euler's theorem,

$$E = \frac{\partial E}{\partial \phi}\, \phi + \frac{\partial E}{\partial v}\, v + \frac{\partial E}{\partial m_1}\, m_1 + \dots$$

Therefore $E = t\phi - pv + \mu_1 m_1 + \mu_2 m_2 + \dots$

Whence
$$
\left.
\begin{array}{l}
I = t\phi + \mu_1 m_1 + \dots, \\
\psi = - pv + \mu_1 m_1 + \dots, \\
\zeta = \mu_1 m_1 + \dots
\end{array}
\right\} \quad \dots (2).
$$

Differentiating the first equation of (2) and using (1) we have

$$0 = \phi dt - v dp + m_1 d\mu_1 + \dots \qquad \dots (3).$$

From the other equations of (2) we have, making use of (3),

$$
\begin{aligned}
dI &= t d\phi + v dp + \mu_1 dm_1 + \dots, \\
d\psi &= - \phi dt - p dv + \mu_1 dm_1 + \dots, \\
d\zeta &= - \phi dt + v dp + \mu_1 dm_1 + \dots.
\end{aligned}
$$

These last four equations correspond to four fundamental equations connecting the variables

$$
\begin{aligned}
&t,\ p,\ \mu_1,\ \mu_2,\ \dots, \\
&I,\ \phi,\ p,\ m_1,\ m_2,\ \dots, \\
&\psi,\ t,\ v,\ m_1,\ m_2,\ \dots, \\
&\zeta,\ t,\ p,\ m_1,\ m_2,\ \dots.
\end{aligned}
$$

104. Heterogeneous mixtures. Between 1875–8, J. Willard Gibbs published a remarkable series of memoirs 'On the equilibrium of heterogeneous substances.' The immense value of his methods and results are now fully recognised by all who study physical chemistry. The 'Phase Rule' alone has served to classify and explain experimental facts which without it would have been a maze of scattered detail.

Gibbs considers a mixture composed of a certain minimum number of independent substances called 'components'; the mixture consists of a number of homogeneous portions called 'phases' into each of which any of the components may enter. Such a mixture, for example, might be a solution of a non-volatile salt in water in contact with the anhydrous salt and with water vapour. There are two 'components'—the salt and water, and there are three 'phases'—the solid salt, the solution, the vapour; each phase being homogeneous.

105. *Conditions of equilibrium at constant temperature and pressure.* Gibbs first considered the case where every component is present in each phase.

Let the number of components be n and the number of phases p.

Let $m_1' \dots m_n'$ be the masses of the component substances $s_1 \dots s_n$, in the first phase.

Let $m_1'' \dots m_n''$ be the masses of the component substances $s_1 \dots s_n$, in the second phase.

Let $m_1^p \dots m_n^p$ be the masses of the component substances $s_1 \dots s_n$, in the pth phase.

Then since

$$d\zeta = - \phi dt + v dp + \mu_1 dm_1 + \mu_2 dm_2 + \dots \quad (\S\ 103)$$

for a given phase (a homogeneous mixture), then at constant temperature and pressure,

$$d\zeta = \mu_1 dm_1 + \mu_2 dm_2 + \dots$$

for a phase; also $\mu_1 = \partial\zeta/\partial m_1$, etc.

In equilibrium, ζ is a minimum at constant temperature and pressure; or $d\zeta = 0$ to the first order.

Therefore

$$\left.\begin{aligned}
0 = \quad & \mu_1' dm_1' + \dots + \mu_n' dm_n' \\
& + \mu_1'' dm_1'' + \dots + \mu_n'' dm_n'' \\
& \dots\dots\dots\dots\dots\dots\dots\dots\dots\dots \\
& + \mu_1^p dm_1^p + \dots + \mu_n^p dm_n^p
\end{aligned}\right\} \quad \dots\dots(1).$$

The total mass of each component is given. Therefore

$$\left.\begin{aligned}
dm_1' + dm_1'' + \dots + dm_1^p = 0 \\
\dots\dots\dots\dots\dots\dots\dots\dots\dots\dots\dots \\
dm_n' + dm_n'' + \dots + dm_n^p = 0
\end{aligned}\right\} \quad \dots\dots(2).$$

Multiplying the equations (2) by $\lambda_1 \ldots \lambda_n$, adding them to (1), and equating to zero the coefficients of the various differentials, we have

$$\mu_1' + \lambda_1 = 0, \; \mu_1'' + \lambda_1 = 0, \text{ etc.}$$
$$\mu_2' + \lambda_2 = 0, \; \mu_2'' + \lambda_2 = 0, \text{ etc.}$$

and so on.

Therefore
$$\left.\begin{array}{l} \mu_1' = \mu_1'' = \ldots = \mu_1{}^p \\ \ldots\ldots\ldots\ldots\ldots\ldots\ldots\ldots \\ \mu_n' = \mu_n'' = \ldots = \mu_n{}^p \end{array}\right\} \ldots\ldots\ldots\ldots(3).$$

Hence the potentials of the substance s_1 are the same in each phase, and so for $s_2, s_3, \ldots s_n$.

106. *The phase rule.* The composition of each phase is determined by the ratio of $m_1 : m_2 : \ldots : m_n$ for the phase; or the composition of each phase is expressed by $(n-1)$ variables.

Hence there are $p\,(n-1)$ variables which determine the composition of the phases; these, together with the temperature and the pressure, make $p\,(n-1) + 2$ variables in all. To determine these there are the equations (3), which are $n\,(p-1)$ in number. Hence there are

$$p\,(n-1) + 2 - n\,(p-1)$$

variables which are arbitrary or $(n - p + 2)$ independent variables.

This result, that the number of independent variables is $(n - p + 2)$, where n is the number of constituents and p the number of phases, constitutes the 'phase rule' of Gibbs. Its great value as a guide in experimental work was first shown by the researches of Bakhuis Roozeboom on heterogeneous chemical systems.

107. *Non-variant system.* This is a system for which $p = n + 2$, so that the number of independent variables or 'degrees of freedom' is 0; the phases can only exist in contact at a definite temperature and pressure, each phase having a definite composition.

Mono-variant system ('condensed' system). In this case, $p = n + 1$ and the number of degrees of freedom is 1; the temperature, *or* the pressure, *or* the composition of one of the phases is arbitrary and the rest are then determinate. Thus at any given temperature, for instance, the pressure and nature of the phases are definite.

Di-variant system. Here $p = n$ and there are 2 degrees of freedom; any two of the temperature, pressure and compositions of the phases are arbitrary and the rest determinate. For instance, at any given temperature and any given pressure, the nature of the phases is definite.

108. *Illustrations of the phase rule.*

(1) *Ice, water, and water vapour in contact.* There is one component substance and there are three phases, solid, liquid and vapour; $n = 1$, $p = 3$. The system is non-variant and the three can only exist in contact at a definite temperature and a definite pressure. (The temperature is − ·0074° C. and the pressure 4·6 mm.) This point is called the 'triple' point.

Using very high pressures, Jammann[1] found three crystalline varieties of ice and Bridgman[2] by extending the range of temperature and pressure has found five. In the latter case $n = 1$, $p = 7$. The phase rule suggests that actually p cannot exceed $n + 2$, otherwise there would be negative degrees of freedom, so that with one component $(n = 1)$, p cannot actually exceed 3. This agrees with the observations of Jammann and Bridgman, who in no case were able to obtain equilibrium between more than three of the possible phases.

(2) *Solution of common salt in water.* (a) The salt is supposed non-volatile and so does not exist in the vapour phase. If the water is in excess, then there are two components (salt, water) and two phases (solution, vapour), or $n = 2$, $p = 2$, a di-variant system. The variables are

[1] *Ann. der Physik*, 1900.   [2] *Proc. Amer. Acad.* 1912.

three, viz. the pressure, temperature, and concentration of the solution (*i.e.* the ratio of the mass of the salt to the mass of the water in the solution). Of these any two are independent, by the phase rule. If, for example, they are the temperature and concentration, then the pressure is determinate; or, a solution of any given concentration and temperature has a definite vapour pressure.

(*b*) As the temperature is lowered, a point is reached at which pure ice separates from the solution. There are now three phases (ice, solution, vapour) or $n = 2$, $p = 3$, a mono-variant system. There is only one independent variable. If, for example, it is taken to be the temperature, the pressure and concentration are determinate. Thus for any given temperature, there is definite concentration of the solution and a definite vapour pressure.

(*c*) At a sufficiently low temperature, the limit of solubility is reached, as the concentration increases owing to the separation of ice, and a mixture of ice and salt crystals, called by Guthrie[1] a 'cryohydrate,' is formed. There are now four phases, ice, salt crystals, solution, vapour; $n = 2$, $p = 4$, a non-variant system, which can only exist at one given temperature and pressure, with a definite concentration of the solution. For common salt and water, this temperature is $- 22°$ C. (cryohydric point), the pressure is ·73 mm. and the solution phase is 36 parts of salt to 100 of water.

(3) *Freezing mixtures.* When ice and salt are mixed, some of the salt dissolves in the water adhering to the ice, so that there is salt, ice, solution, and vapour in contact. This system cannot be in equilibrium except at $- 22°$ C.; therefore more ice melts and the system cools until this temperature is reached.

(4) *Water and steam.* Here $n = 1$, $p = 2$; the system is mono-variant. The composition of each phase is definite; one is water only, the other steam only. The variables are

[1] *Phil. Mag.* 1875.

the temperature and pressure, and one of these is independent. Thus for any given temperature there is a definite
vapour pressure.

109. *Thermodynamic surface for a homogeneous mixture
of two components.* One of the five types of fundamental
equation given in § 103 for a homogeneous mixture was one
connecting the variables ζ, t, p, m_1, ... m_n.

Thus for a two-component system $\zeta = F(t, p, m_1, m_2)$ is
a fundamental equation. If ζ refers to unit mass of the
mixture, $m_1 + m_2 = 1$. Denoting m_1 by c,

$$\zeta = F(t, p, c, 1 - c) \text{ or } \zeta = f(t, p, c).$$

The variable c is the 'concentration' of the first substance
in the mixture and may vary from 0 to 1.

For experiments conducted at atmospheric pressure, p is
constant, so that $\zeta = f(t, c)$. If the values of c, t, ζ are
taken as coordinates x, y, z, this relation becomes

$$z = f(y, x),$$

a surface which determines the properties of the mixture
at atmospheric pressure.

A section of this surface by the plane $t = t_0$ is a curve
whose projection on the plane $t = 0$ is of the form $\zeta = F(c)$.
By experiments conducted at temperature t_0 this relation
between ζ and c can be found and the form of this section
determined. The surface can thus be built up from observations made over a range of temperatures.

110. *Distillation of a mixture of two liquids.* If the
liquids mix so as to form a homogeneous mixture, then in
the process of distillation there are two phases (liquid and
vapour) and two components.

Let the ζ-c curve for the liquid phase at temperature
t be found and also the ζ-c curve for the vapour phase
at that temperature. Draw the common tangent touching
the curves at L, V. When the two phases are in equilibrium the potentials of a given component are the same

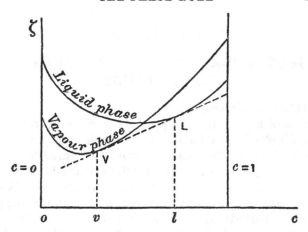

in both phases (§ 105). But the potential $\mu = \partial\zeta/\partial m$; therefore $\partial\zeta/\partial m_1$ has the same value in both phases; so for $\partial\zeta/\partial m_2$. But $m_1 = c$, $m_2 = 1 - c$. Hence it suffices that $d\zeta/dc$ should be the same for both phases. But $d\zeta/dc$ is the same for the two curves at L and V, being the gradient of the tangent; therefore L, V represent co-existent states of the two phases. The abscissae of L, V give the corresponding concentrations, thus determining the composition of the liquid and vapour phases at temperature t.

This method, developed by Roozeboom[1] and applied by him to investigate solid solutions (alloys), is ultimately traceable to Gibbs[2].

[1] *Zeit. phys. Chem.* 1899.
[2] *Trans. Connecticut Acad.* 1875; *Coll. Papers,* vol. I. p. 120.

CHAPTER XI

OSMOTIC PRESSURE; VAPOUR PRESSURE; GAS MIXTURES

111. *Osmotic pressure.*

Cells of plants have the property of allowing water but not dissolved material to pass through their walls. An artificial cell of this character can be prepared by filling a porous cylinder with copper sulphate solution and immersing it in a solution of potassium ferrocyanide. At the middle of the wall a fine membrane of precipitated copper ferrocyanide is deposited. Such a membrane, which behaves like the cell wall of a plant, is called a 'semi-permeable' membrane.

In Pfeffer's experiments the cell, closed at the top and connected to a manometer, was filled with a solution of cane sugar, and then immersed in water kept at a constant temperature. Water entered the cell from outside, but no sugar could escape, and the pressure inside the cell rose to a maximum value, the 'osmotic pressure' of the solution.

112. *Osmotic pressure and gas pressure.*

Pfeffer found that the osmotic pressure is proportional to the concentration of the solution or to the number of gram-molecules of sugar per unit volume; and also that it varies approximately as the temperature. Now for a perfect gas $p = nRt$, where n is the number of gram-molecules of gas per unit volume and t is the temperature, so that the laws of osmotic pressure are the same as the laws of gas pressure. Van 't Hoff, assuming the relation $p = nRt$ for osmotic pressure, proceeded to calculate R from Pfeffer's data for cane sugar.

At temperature 280° abs. a 1 per cent. solution of cane sugar $C_{12}H_{22}O_{11}$ (molecular weight 342) has an osmotic pressure of 505 mm. of mercury.

$\therefore\; p = \frac{505}{760} \times 1013600$ dynes per sq. cm.

$n =$ number of gram-molecules of sugar per c.c. of solution

$\qquad = \frac{1}{100} \times \frac{1}{342}.$ $t = 280.$

$\therefore\; R = \dfrac{505 \times 1013600 \times 34200}{760 \times 280} = 84{\cdot}1 \times 10^{6}.$

For a gas, the constant $R = 83{\cdot}6 \times 10^{6}$.

This remarkable coincidence led to Van 't Hoff's[1] conclusion that the molecules of sugar in a dilute solution produce the same pressure as they would if they were gaseous molecules in the same space.

Van 't Hoff generalised this into a principle true of all finely dispersed material, where the groups of molecules are so far separated that for the greater part of their motion they are beyond each other's sphere of action. (A reasoned account of the foundations of osmotics is given in a paper by Larmor, *Phil. Trans.* 1897.)

113. *Osmotic pressure of gases.*

The following experiment proposed by Arrhenius[2] and carried out by Ramsay[3] throws much light on the osmotic pressure of solutions. H and

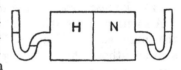

N are equal vessels with a palladium wall between them; one is filled with hydrogen and the other with nitrogen, both at atmospheric pressure. When raised to a high temperature the palladium acts as a semi-permeable membrane permitting the passage of the hydrogen but not that of the nitrogen. When equilibrium is established and the vessels allowed to cool, the vessel H is full of hydrogen at $\frac{1}{2}$ atmosphere pressure and N contains a mixture of nitrogen and hydrogen at $1\frac{1}{2}$ atmospheres pressure.

Now if n was the number of molecules per unit volume in H and N at the beginning (they would be the same, by

[1] *Phil. Mag.* 1888.   [2] *Zeit. phys. Chem.* 1889.   [3] *Phil. Mag.* 1894.

Avogadro's law) and n_1, n_2, n were the numbers of molecules per unit volume of hydrogen in H, of hydrogen in N, and of nitrogen in N after the experiment, then

$$1 \text{ atmosphere } = nRt,$$
$$\tfrac{1}{2} \text{ atmosphere } = n_1 Rt,$$

and
$$1\tfrac{1}{2} \text{ atmospheres } = (n + n_2)\, Rt.$$

$$\therefore \frac{n}{2} = \frac{n_1}{1} = \frac{n + n_2}{3}.$$

$$\therefore n_1 = \tfrac{1}{2}n \text{ and } n_2 = \tfrac{1}{2}n.$$

Thus equilibrium is established when the number of molecules per unit volume of the permeating gas is the same on the two sides of the semi-permeable diaphragm.

Again, the pressure due to the nitrogen after the experiment $= nRt = 1$ atmosphere. Thus the difference of pressure in H and N is the osmotic pressure of the nitrogen. The hydrogen and nitrogen here correspond to the water and cane sugar, and the pressure of the nitrogen to the osmotic pressure in Pfeffer's experiment.

114. *Vapour pressure and osmotic pressure.*

A solution of a salt has a lower vapour pressure than the pure solvent. Van 't Hoff[1] showed that the lowering of vapour pressure was proportional to the osmotic pressure of the solution. Arrhenius[2] has obtained this result in the following manner.

The long tube closed by a semi-permeable membrane A contains sugar solution. The lower end is immersed in a vessel of water and the whole is placed under a cover from which the air is exhausted.

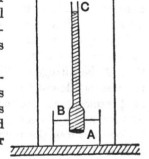

[1] *Zeit. phys. Chem.* 1887.   [2] *Zeit. phys. Chem.* 1889.

Let p be the vapour pressure of water,

   p' be the vapour pressure of the solution,

and P be the osmotic pressure of the solution.

The pressure at B is p and this is the pressure on the lower side of A. On the upper side of A in the solution the pressure is $p + P$. The pressure in the tube at C is p'.

From the equilibrium of the column of solution AC, $p + P - p' = \rho h$, where ρ is the weight per unit volume of the solution and h is the height of C above B.

From the equilibrium of the column of vapour between the levels of B and C, $p - p' = \sigma h$, where σ is the weight per unit volume of the vapour, assumed constant between B and C.

$$\therefore \frac{p - p' + P}{p - p'} = \frac{\rho}{\sigma}$$

or

$$\frac{P}{p - p'} = \frac{\rho}{\sigma} - 1$$

or, as ρ/σ is large, $\dfrac{P}{p - p'} = \dfrac{\rho}{\sigma}$ very approximately,

or

$$p - p' = \frac{\sigma}{\rho} P.$$

115. Vapour pressure and concentration.

$$\frac{p - p'}{p} = \frac{\sigma}{\rho} \frac{P}{p}.$$

Now $p = nRt$ and $P = n'Rt$ where n is the number of molecules of vapour per unit volume and n' is the number of molecules of sugar per unit volume.

$$\therefore \frac{p - p'}{p} = \frac{\sigma n'}{\rho n}.$$

Now if n'' is the number of molecules of water per unit volume, $\dfrac{\sigma}{\rho} = \dfrac{n}{n''}$, assuming the density σ of the dilute solution equal to that of water.

[The equation $\dfrac{\sigma}{\rho} = \dfrac{n}{n''}$ implies that the solvent (in the argument, water) has the same molecular weight in the liquid and vapour states.]

$$\therefore \frac{p - p'}{p} = \frac{n'}{n''} = \frac{\text{number of molecules of sugar}}{\text{number of molecules of water}},$$

per unit volume.

$$\therefore \frac{p - p'}{p} = \frac{1}{N},$$

where N is the number of molecules of water in which 1 molecule of sugar is dissolved.

Thus the 'relative' lowering of the vapour pressure is independent of the nature of the solvent or the substance dissolved, and depends only upon the relative number of their molecules.

This result was first obtained experimentally by Raoult[1] for fourteen different solvents and with various dissolved substances; the theoretical deduction is due to Van 't Hoff and Arrhenius.

116. *Boiling point of a solution.*

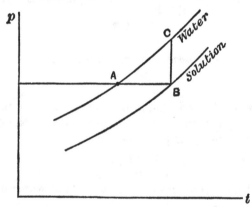

The upper curve shows the relation between pressure and temperature when water is at the boiling point; the lower curve indicates it for the solution. The ordinate of A is the atmospheric pressure p and the

----

[1] *Zeit. phys. Chem.* 1888.

abscissa t the boiling point of water. The abscissa $t + dt$ of B is the boiling point of the solution at atmospheric pressure, so that dt is the rise of the boiling point due to the presence of the dissolved substance. BC is the lowering of the vapour pressure due to the presence of dissolved substance and is equal to p/N, in the notation of the preceding paragraph.

But $\dfrac{BC}{BA} = \dfrac{dp}{dt}$ for water at the point of vaporisation, and by Clapeyron's equation

$$\frac{dp}{dt} = \frac{L}{(v - v')\,t},$$

where L is the latent heat, and v, v' refer to the vapour and the liquid (water); hence since v'/v is negligible,

$$\frac{BC}{BA} = \frac{L}{vt}.$$

$$\therefore \frac{p}{Ndt} = \frac{L}{vt}.$$

$$\therefore dt = \frac{pvt}{NL}.$$

Using $pv = Rt$ for the vapour,

$$dt = \frac{Rt^2}{NL}.$$

If L is the latent heat of vaporisation of 1 mol of water, then $R = 2$ calories, approx. (§ 5), so that

$$dt = \frac{2t^2}{NL}.$$

Thus for solutions containing n mols of the dissolved substance to 100 mols of the solvent,

$$dt = \frac{2t^2}{L}\,\frac{n}{100},$$

$$dt = (\cdot 02)\frac{t^2 n}{L}.$$

This formula gives the rise of the boiling point and has been verified experimentally by Beckmann[1] for some ten different solvents.

117. *Determination of molecular weight of a dissolved substance from the rise of the boiling point.*

A numerical case will suffice.

It is found that if 1 gram of iodine is dissolved in 50 grams of ether, the boiling point is raised by ·167 of a degree.

For ether, the boiling point is $34°·9$ and the latent heat per gram is $90·45$. The molecular weight of ether, $(C_2H_5)_2O$, is 74.

$$\therefore \; L = 90·45 \times 74.$$

$$\therefore \; \frac{·02t^2}{L} = \frac{1}{50} \times \frac{(307·9)^2}{90·45 \times 74} = ·283.$$

Therefore for ether, $\qquad dt = n \,(·283).$

Now if M is the molecular weight of the dissolved iodine, $1/M$ is the number of mols of iodine in the solution; also $50/74$ is the number of mols of ether.

$$\therefore \; \frac{1}{M} \Big/ \frac{50}{74} = n/100.$$

$$\therefore \; M = \frac{148}{n} = \frac{148 \times ·283}{dt}$$

and $\qquad\qquad dt = ·167.$

$$\therefore \; M = \frac{148 \times ·283}{·167}$$

$$= 251.$$

The atomic weight of iodine is 127, so that in an ether solution the molecule consists of two atoms.

Observations on solutions of phosphorus and sulphur in carbon disulphide show that in the solution the phosphorus molecule contains four atoms and the sulphur molecule eight atoms.

---

[1] *Zeit. phys. Chem.* 1889.

118. *Freezing point of solutions.*

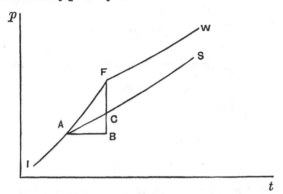

FW is the vapour pressure curve for water,

and   FI   „   „   „   ice.

F is the triple point for the equilibrium of ice, water, and vapour at the same pressure, 4·6 mm. {§ 108 (1).}

AS is the vapour pressure curve for the solution.

When the solution freezes, it is in equilibrium with the ice that freezes out and with the vapour, so that all three are at the same pressure. Therefore A is the freezing point of the solution. The lowering of the freezing point is therefore AB (= dt).

Let L = the latent heat of vaporisation of 1 mol of water.

L' = the latent heat of fusion of 1 mol of ice.

v = the volume of 1 mol of water vapour.

p = vapour pressure at F or A (approx.).

Now $\dfrac{FB}{dt} = \dfrac{dp}{dt}$ for the ice-vapour curve

$$= \frac{L + L'}{vt}$$, by Clapeyron's equation, for $L + L'$

is the latent heat of *vaporisation* of 1 mol of ice, and the volume of 1 mol of ice is neglected in comparison with v the volume of 1 mol of water vapour.

$$\therefore \frac{FB}{dt} = \frac{(L + L')\,p}{Rt^2} \;,\; \text{using } pv = Rt \text{ for the vapour.}$$

So $\dfrac{CB}{dt} = \dfrac{dp}{dt}$ for the solution-vapour curve

$$= \frac{dp}{dt} \text{ for the water-vapour curve, approx., since}$$

the solution is dilute,

$$= \frac{Lp}{Rt^2}.$$

From these we have

$$FC = \frac{L'p\,dt}{Rt^2}.$$

But FC is the lowering of the vapour pressure of the solution and $= p/N$.

$$\therefore \frac{1}{N} = \frac{L'\,dt}{Rt^2}.$$

$$\therefore dt = \frac{Rt^2}{NL'} = \frac{2t^2}{NL'}.$$

Hence for a solution of n mols of a substance in 100 mols of solvent,

$$dt = \frac{(\cdot 02)\,nt^2}{L'},$$

where L' is the latent heat of *fusion* of the solvent.

The measurement of the depression of the freezing point leads to the determination of the molecular weight of the dissolved substance.

These methods of determining molecular weights have been of the greatest value in the study of the molecule, because before they were discovered, the determination of molecular weights was limited to cases where the substance could be vaporised without chemical change.

119. *Gas mixtures and dilute solutions.*

In a gas mixture, Dalton showed that the pressure is the same as if each constituent exerted the pressure it would if it alone filled the whole volume. This property

is true also of the entropy. The entropy of the mixture
is the sum of the entropies each would have if at the
same temperature it occupied the whole volume of the
mixture. This result is due to Gibbs[1]. It has been proved
by Planck[2] in the following way, by the use of a semi-
permeable membrane, a method which has become an
important feature of many theoretical deductions in
physical chemistry.

The cylinder shown in the figure
has four pistons, of which A, B are
fixed and A', B' are movable but
with $A'B'$ constant and equal to
AB. The nature of the system is
described in the figure.

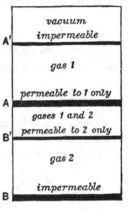

At first let A', B' coincide with A,
B, so that the space AB contains
the mixture of gases 1 and 2. The
connected pistons A', B' are slowly
raised so that gas 1 passes into the
space AA' and gas 2 into the space
BB' through the semi-permeable
pistons A, B'.

Complete separation of the two gases is effected when
B' comes into contact with A.

To find the work done in this process, we find the
pressures in the three spaces at any stage. By the conditions of osmotic equilibrium for gases (§ 113) the number
of molecules per unit volume of the gas 1 is the same (n_1)
on each side of A and the number per unit volume of the
gas 2 is (n_2) on each side of B'.

Therefore the pressures in $A'A$, AB', $B'B$ are p_1, p, p_2
where $$p_1 = n_1 Rt, \quad p = (n_1 + n_2) Rt, \quad p_2 = n_2 Rt.$$

The total upward thrust on the combined pistons A', B'
(each taken of unit area) is

$$p_1 - p + p_2 = n_1 Rt - (n_1 + n_2) Rt + n_2 Rt = 0.$$

[1] *Coll. Papers*, vol. I. p. 156.   [2] *Wied. Ann.* 1883.

Hence no work is done in this isothermal process. The internal energy of the gases is a function of the temperature only and is therefore constant. Now

$$dE = td\phi - dW$$

and since $dE = 0$ and $dW = 0$, $d\phi = 0$ or the entropy of the system is unaltered by the process. But at the end, each gas has the same volume as the initial volume of the mixture, so that the theorem follows.

120. *The entropy of a gas mixture.*

It has been shown (§ 39) that for a perfect gas

$$\phi = c_v \log t + R \log v + C, \text{ per mol, where } R = 83 \cdot 6 \times 10^6$$

$$= c_v \log t + R \log \left(\frac{t}{p}\right) + \kappa.$$

Consider a mixture of n_1 mols of a first gas, n_2 of a second, etc.

By the theorem of the preceding article, for the mixture

$$\phi = n_1 \left[c_{v1} \log t + R \log \frac{t}{p_1} + \kappa_1 \right]$$
$$+ n_2 \left[c_{v2} \log t + R \log \frac{t}{p_2} + \kappa_2 \right]$$
$$+ \text{etc.},$$

where $p_1, p_2, \ldots$ are the pressures each gas would exert if it occupied the whole volume, *i.e.* the 'partial' pressures of Dalton.

Now $p_1 = n_1 Rt$, $p_2 = n_2 Rt$, etc. and the pressure p of the mixture is $(n_1 + n_2 + \ldots) Rt$.

$$\therefore \ p_1 = \frac{n_1}{\Sigma n} p, \text{ etc. or } p_1 = c_1 p, \ p_2 = c_2 p, \text{ etc.}$$

where $\quad c_1 = \frac{n_1}{\Sigma n}$, the 'concentration' of the gas 1, etc.

$$\therefore \ \phi = \Sigma n_1 \left[c_{v1} \log t + R \log \left(\frac{t}{c_1 p}\right) + \kappa_1 \right] \ \ldots\ldots(1)$$

for the mixture.

121. *Increase of entropy due to diffusion.*
Before mixture, the total entropy was

$$\Sigma n_1 \left(c_{v1} \log t + R \log \frac{t}{p} + \kappa_1\right) \quad \dots\dots(2).$$

Thus the increase of entropy due to their diffusion into a uniform mixture is the excess of (1) over (2), which is

$$\Sigma \left(- n_1 R \log c_1\right)$$
$$= \Sigma n_1 R \log \left(\frac{n_1 + n_2 + \dots}{n_1}\right),$$

a positive quantity. Thus the increase of entropy is independent of the nature of the gases and depends only upon the numbers of molecules present.

122. *Thermodynamics of a gas mixture. Dissociation.*
Planck[1] constructed the ζ function for a gas mixture.
For a mol of gas the internal energy $= c_v t + h$, where h is a constant.
Therefore for the mixture

$$E = \Sigma n_1 \left(c_{v1} t + h_1\right)$$
and $\quad \phi = \Sigma n_1 \left[c_{v1} \log t + R \log \frac{t}{c_1 p} + \kappa_1\right]$
and $\quad pv = (n_1 + n_2 + \dots) Rt.$

$$\therefore \ \zeta = E - t\phi + pv$$
$$= \Sigma n_1 \left(c_{v1} t + h_1\right) - t\Sigma n_1 \left[c_{v1} \log t + R \log \frac{t}{c_1 p} + \kappa_1\right] + Rt\Sigma n$$

or $\quad \zeta = \Sigma n_1 \left[c_{v1} (t - t \log t) + h_1 - Rt \log \frac{t}{p}\right.$
$$\left. - \kappa_1 t + Rt + Rt \log c_1\right]$$
$$= \Sigma n_1 [\phi_1 + Rt \log c_1],$$

where each ϕ is a function of t, p only.
Thus ζ is expressed as a function of $t, p, n_1, n_2, \dots$.
This is a fundamental equation of the Gibbs $\zeta, t, p,$ $m_1, m_2, \dots$ type (§ 103) from which all the thermodynamic

[1] *Wied. Ann.* 1883.

properties of the homogeneous gas mixture can be found. Planck proceeds to use the ζ function, with its minimum property, to find a formula from which the state of chemical dissociation of a gaseous substance can be determined for any given temperature and pressure.

123. *Thermodynamics of dilute solutions.*

Planck[1] found a ζ function of the same form for dilute solutions. By its use he deduced Henry's law that the concentration of a dissolved gas is proportional to the pressure of the gas on the solution; the formulae for the change of the boiling point and freezing point in terms of number of molecules of dissolved substance per molecule of the solvent; and the solution of a number of problems on the state of ionisation of dissolved substances.

124. *Surface tension and temperature.*

The formula $T = T_0 (1 - a\theta)^n$ has been shown by experiments on a number of organic liquids to represent the variation of surface tension T of a liquid in contact with air and its vapour at temperature $\theta°$ C. The numbers a, n are constant for a given liquid. For the liquids examined a is of order 10^{-3}, while n is nearly constant for all the liquids, and about equal to $1 \cdot 2$.

At the critical temperature, where there is no distinction between liquid and vapour, the surface tension must be zero, so that the critical temperature θ_c is given by $1 - a\theta_c = 0$ or $\theta_c = 1/a$. There is very close agreement between critical temperatures found by surface tension experiments which determine a and thence θ_c, and critical temperatures observed directly.

The Thomson equation may now be used to connect the energy per unit area of the surface between a liquid and its vapour (E) with the surface tension. Let the area of a given portion of surface increase from A to $A + dA$. Then

[1] *Wied. Ann.* 1887.

the work done by the surface tension $= \Sigma \,(Tds)\, dn$ taken round the perimeter

$$= T\Sigma dsdn$$
$$= TdA$$

and the increase of energy is EdA.

The temperature being constant, the increase of $\psi =$ the work done $= TdA$ and the increase of $E = EdA$.

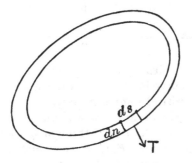

The Thomson equation is

$$\psi = E + t \left(\frac{d\psi}{dt}\right)_v \,,$$

where ψ, E may be changes in ψ, E, the equation being linear.

$$\therefore \; TdA = EdA + t\left[\frac{d}{dt}\,(TdA)\right]_v \cdot$$

$$\therefore \; T = E + t\frac{dT}{dt}\,,$$

an equation connecting surface tension and surface energy at temperature t.

Now let the curve for T, t be drawn, using

$$T = T_0\,(1-a\theta)^n \text{ and } \theta = t - 273 \cdot 1.$$

If the tangent at any point P meets OT in Q (p. 122),

$$OQ = PN - LQ = PN - LP\tan\phi$$
$$= T - t\frac{dT}{dt} = E.$$

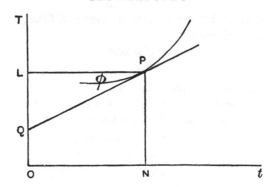

Thus PN denotes T and OQ denotes E.

The actual curve for T, t is shown below, and touches the t axis at the critical temperature t_c.

If the tangent at any point P meets the T axis in Q and a parallel to the t axis through Q meets NP in P', P' is a point on the (E, t) curve, which is thus readily drawn, and is indicated by the dotted line.

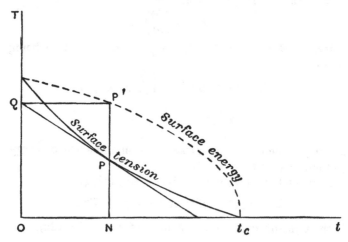

CHAPTER XII

THERMOELECTRIC PHENOMENA

125. *The Seebeck, Peltier, and Thomson effects.*

In 1821 Seebeck discovered that if the ends of two wires of different metals are joined so as to form a closed circuit and the junctions are maintained at different temperatures, an electric current passes round the circuit.

In 1834 Peltier found that when a current passes through the junction of two wires of different metals, heat is evolved (or absorbed) at the junction.

[The heat given out per second is proportional to the current and is Πi, where Π is the Peltier coefficient for the two metals at temperature t.]

Both of these effects are reversible.

Thus the discovery of Peltier shows that in the thermoelectric circuit of Seebeck heat is absorbed at the hot junction and given out at the cold one; it appears that the circuit develops electric power from the heat reversibly like a Carnot engine.

When Thomson[1] in 1851 applied the laws of thermodynamics to the thermoelectric circuit, he found that if the only thermal effect were the Peltier effect at the junctions, then the electromotive force in the circuit would be subject to the same law of variation with the temperatures of the two junctions whatever be the metals of which it is composed. This result being at variance with known facts, Thomson concluded that 'an electric current produces different thermal effects, according as it passes from hot to cold, or cold to hot, in the same metal.'[2]

This evolution or absorption of heat when a current passes along a wire whose temperature varies from point to point, is the Thomson 'effect.' If P and P' are consecutive cross-sections of a wire at temperatures t and $t + dt$, and a current

---

[1] Thomson, Sir W., *Coll. Papers*, vol. I. p. 232 *et seq.* 1854.
[2] *Ibid.* p. 319.

i is flowing from P towards P', the heat absorbed per second from PP' is $\sigma i dt$, where σ is the Thomson coefficient of the wire at P; for a given metal, σ is a function of the temperature t only, for if it depended on the cross-section of the wire or any other such variable a current might be maintained by the application of heat to a homogeneous metallic wire.

Thomson called σ the 'specific heat of electricity in the given metal,' for it is the heat absorbed by unit current through a rise of temperature dt equal to unity.

Since the Peltier and Thomson reversible effects vary directly as the current and the usual Joule irreversible heat effect varies as the square of the current, the latter is small compared with the former for the small currents of the thermoelectric circuit.

126. *Theory of the thermoelectric circuit.*

a, b are wires of two different metals, t_1, t_2 the temperatures of the junctions $(t_1 > t_2)$.

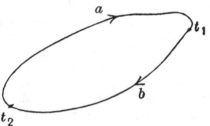

Let V be the electromotive force and let a small current i pass round the circuit. Then (all the quantities being estimated per second) the work done is Vi and the corresponding heat comes from the Peltier and Thomson effects.

Owing to the former, heat $\Pi_1 i$ is taken in at the hot junction and $\Pi_2 i$ given out at the cold one, where Π_1, Π_2 are the Peltier coefficients at temperatures t_1, t_2 for the two metals.

Owing to the latter, the heat absorbed in a is $\displaystyle\int_{t_2}^{t_1} \sigma_a\, i dt$ and the heat given out in b is $\displaystyle\int_{t_2}^{t_1} \sigma_b\, i dt.$

By the first law,

$$V = \Pi_1 - \Pi_2 + \int_{t_2}^{t_1} (\sigma_a - \sigma_b)\, dt$$

and by the second law, $\int \dfrac{dQ}{t} = 0$ for the cycle, so that

$$\frac{\Pi_1}{t_1} - \frac{\Pi_2}{t_2} + \int_{t_2}^{t_1} \frac{\sigma_a - \sigma_b}{t}\, dt = 0.$$

If these equations are applied to a thermoelectric pair, whose junctions are at temperatures $t, t + dt$, and if dV is the electromotive force, they become

$$\left. \begin{array}{l} dV = d\Pi + (\sigma_a - \sigma_b)\, dt \\[2mm] d\left(\dfrac{\Pi}{t}\right) + \dfrac{\sigma_a - \sigma_b}{t}\, dt = 0 \end{array} \right\}.$$

and

$$\left. \begin{array}{l} \therefore\ \dfrac{dV}{dt} = \dfrac{d\Pi}{dt} + \sigma_a - \sigma_b \\[2mm] t\dfrac{d}{dt}\left(\dfrac{\Pi}{t}\right) + \sigma_a - \sigma_b = 0 \end{array} \right\}.$$

$$\therefore\ t\frac{d}{dt}\left(\frac{\Pi}{t}\right) + \frac{dV}{dt} - \frac{d\Pi}{dt} = 0$$

or

$$\frac{\Pi}{t} = \frac{dV}{dt},$$

or

$$\Pi = t\frac{dV}{dt}.$$

$$\therefore\ \sigma_a - \sigma_b = -t\frac{d^2V}{dt^2}.$$

Thomson's conjecture, based on theoretical reasoning, that there must be reversible heat effects in the thermoelectric circuit other than the Peltier heat effect, thus led to the formula

$$\Pi = t\frac{dV}{dt};$$

he says 'it would be important to test the expression by direct experiment and so confirm the doubtful part of the

theory,' and he proceeds to indicate the lines on which
the experiments should be carried out[1]. The formula has
since been amply verified.

127. *Reversible cell.*

In the Volta cell, the elements are zinc/dilute sul-
phuric acid/copper. In the normal action of the cell zinc
passes into solution and hydrogen separates at the copper
pole; the current passes through the cell from the zinc to
the copper. If a current is passed through the cell in the
opposite direction from some external source, copper is
dissolved and hydrogen liberated at the zinc pole.

The original state is not restored when the current is
reversed; the cell is of an irreversible type.

In the Daniell cell, the elements are zinc/zinc sulphate
solution/copper sulphate solution/copper. Zinc is dissolved in dilute acid and forms zinc sulphate; copper is
deposited from the copper sulphate solution; and the
current passes through the cell from the zinc to the copper.

When a current is passed through the cell in the opposite
direction the deposited copper redissolves and the zinc is
deposited, so that the original state of the cell is restored;
the cell is of a reversible type.

128. *The* E.M.F. *of a Daniell cell.*

In 1851, Thomson[2] calculated the E.M.F. of a Daniell
cell on the assumption that the work done by the cell in
maintaining the current is equal to the heat produced by
the chemical change, *i.e.* by the solution of zinc and the
deposition of an equivalent amount of copper. The relation between the current and the amount of chemical
change was known from the data of electrolysis, and the
heat developed in the chemical change was known by
calorimetry. These were sufficient to give the value 1·086
volts as the E.M.F., which agreed closely with observed
values.

[1] *Coll. Papers,* vol. I. p. 250.
[2] *Ibid.* p. 479.

In this calculation there is just the same error as that made by Mayer in his determination of the mechanical equivalent of heat by equating the work done in compressing a gas to the heat evolved, and the error was about as difficult to detect; in both cases the possibility of heat being developed from the internal energy of the system was overlooked.

It is of interest to recall that shortly after that time, Thomson (with Joule) had begun the series of experiments which cleared up the error Mayer had made; and that in his work on the thermoelectric circuit he had even deduced the necessity for local heat effects (the Thomson effect) in the wires, of the kind which he had omitted in his theory of the Daniell cell. He must have known it all then. It was not however until some thirty years afterwards that Helmholtz[1] gave the corrected theory, which follows.

129. *Theory of a reversible cell.*

Helmholtz's formula was obtained by the use of a Carnot cycle; it can also be directly deduced from the Thomson equation

$$\psi = E + t \left(\frac{\partial \psi}{\partial t} \right)_v .$$

(In this equation differences of ψ and E in any two states may be used as values of ψ, E, as the equation is linear.) Let a charge e pass through the cell and V be the E.M.F. of the cell. The work done by the cell is Ve and this is equal to the diminution of ψ, at constant volume.

$$\therefore \ Ve = E + t \frac{\partial}{\partial t} (eV)_v$$

$$Ve = E + te \left(\frac{\partial V}{\partial t} \right)_v .$$

E, the decrease of the internal energy, is the same as if the final state of the cell had been produced by direct chemical action. Therefore if λ is the heat of reaction of

---

[1] *Sitzungsber. Berl. Akad.* 1882.

the chemical changes which occur during the passage of unit charge at constant temperature, then $E = \lambda e$.

$$\therefore \ V = \lambda + t \left(\frac{\partial V}{\partial t}\right)_v \text{ (Helmholtz).}$$

Thomson assumed $V = \lambda$, and since in the Daniell cell, $\left(\frac{\partial V}{\partial t}\right)_v$ is small, he found a result agreeing with the known value of the E.M.F.

Helmholtz's formula has been verified by the experiments of Jahn[1]. It should be noticed that the volume of the system has been supposed constant, which is approximately true for cells in which gases are not evolved in the chemical changes.

If Π is the sum of the Peltier effects at the junctions of the cell $\Pi = t \dfrac{\partial V}{\partial t}$ (§ 126), so that Helmholtz's formula becomes $V = \lambda + \Pi$. This too was verified by Jahn[2].

---

[1] *Wied. Ann.* 1886.   [2] *Wied. Ann.* 1888 and 1893.

CHAPTER XIII

GAS THEORY AND VARIATION OF SPECIFIC HEAT WITH TEMPERATURE

130. *Elements of gas theory.*

A gas consists of a large number of molecules moving with large velocities. For the greater part of its path, each molecule is not acted on by any sensible force and the paths are straight lines. When two molecules 'encounter' each other, *i.e.* come sufficiently near for their mutual action to be appreciable, they swerve round one another and their directions of motion change. The straight path between any two encounters is the 'free path' of the molecule.

The velocity of any one molecule is very irregular, but if a statistical view and not an individual one is adopted and the molecules are distributed into groups according to their velocity, a regularity appears. Maxwell[1] showed that the number of molecules whose velocities lie between u, v, w and $u + du, v + dv, w + dw$ is proportional to

$$e^{-a(u^2+v^2+w^2)}\,du\,dv\,dw.$$

131. *Equipartition of energy.*

The ultimate result of encounters between molecules was shown by Maxwell[2] and others to lead to the equipartition of energy between the different effective coordinates (degrees of freedom) of the system, on the Newtonian theory of dynamics.

132. *Gas pressure.*

This is due to the impacts of the molecules on the walls of the containing vessel.

[1] *Collected Works*, vol. I. p. 377, and Jeans' *Dynamical Theory of Gases*, chap. II.
[2] *Collected Works*, vol. II. p. 713, and Jeans, chap. V.

BT

9

Let the gas be contained in a rectangular box, with the axes of x, y, z normal to its faces, and suppose the molecules to be perfectly elastic point spheres.

Consider a face of the box normal to Ox; if u, v, w are the velocities of a molecule on impact, its velocities after impact are $-u$, v, w and the impulse on the wall $= 2mu$, where m is the mass of a molecule.

Let n be the number of molecules per unit volume whose velocity lies between u, v, w and $u + du$, $v + dv$, $w + dw$. Since the gas is at rest, half the molecules are moving towards the face and half away. Hence the number of molecules which impinge on an area dS in time dt is $u\,dt\,dS\,\dfrac{n}{2}$, and the impulse communicated is $u\,dt\,dS\,\dfrac{n}{2}\,2mu$ or mnu^2dSdt. Hence the force on dS is mnu^2dS or the pressure on the face is mnu^2. For all the molecules the pressure is

$$\Sigma\,mnu^2 = m\Sigma nu^2.$$

If $\overline{(u^2)}$ is the mean value of u^2, per unit volume,

$$\overline{(u^2)} = \frac{\Sigma nu^2}{\Sigma n}$$

or $\Sigma nu^2 = N\,\overline{(u^2)}$ where N is the number of molecules per unit volume.

Therefore the pressure p, on the face normal to Ox, is $mN\,\overline{(u^2)}$. So for the other faces normal to Oy and Oz,

$$p = mN\,\overline{(v^2)} \quad \text{and} \quad p = mN\,\overline{(w^2)}.$$

$$\therefore\quad \overline{(u^2)} = \overline{(v^2)} = \overline{(w^2)}$$

and
$$3p = mN\,[\overline{(u^2)} + \overline{(v^2)} + \overline{(w^2)}].$$

If V is the velocity of a molecule, $V^2 = u^2 + v^2 + w^2$.

$$\therefore\quad \overline{(V^2)} = \overline{(u^2)} + \overline{(v^2)} + \overline{(w^2)}.$$

$$\therefore\quad p = \tfrac{1}{3}mNC^2, \quad \text{where} \quad C = \sqrt{\overline{(V^2)}},$$

(Jeans, chap. VI), or the pressure is $\tfrac{2}{3}$ of the kinetic energy of the molecules per unit volume.

If ρ is the mass per unit volume or density of the gas, $\rho = mN$.

Therefore $\qquad p = \tfrac{1}{3}\rho C^2,$

or $\qquad\qquad pv = \tfrac{1}{3}C^2,$

where v is the volume of unit mass. C is the 'velocity of mean square'; it is the velocity of a molecule whose energy is equal to the average kinetic energy of the molecules.

133. The velocity $C = \sqrt{\dfrac{3p}{\rho}}$, and is readily calculated for a given gas. For oxygen at $0°$ C. and atmospheric pressure,

$\rho = \cdot001429$ gr. per c.c. and $p = 1013600$ dynes per sq.cm.

$$\therefore\; C = \sqrt{\frac{3 \times 1013600}{\cdot001429}}$$

$$= 46100 \text{ cm. per second.}$$

134. *Boyle's law.*

Since the heat of a gas on the kinetic theory is the energy of the molecules, it will be assumed that C is constant for constant temperature.

Therefore $pv = $ constant at constant temperature (Boyle's law).

135. *Temperature.*

This is defined in the kinetic theory by the equations

$$m\,\overline{(u^2)} = m\,\overline{(v^2)} = m\,\overline{(w^2)} = Rt,$$

where R is a constant.

Therefore $\qquad mC^2 = 3Rt,$

and $\qquad\qquad p = \tfrac{1}{3}mNC^2 = NRt.$

Thus if R is taken as $13\cdot8 \times 10^{-17}$, the temperature thus defined is Thomson's absolute temperature.

Also since

$$\tfrac{1}{2}m\,\overline{(u^2)} = \tfrac{1}{2}m\,\overline{(v^2)} = \tfrac{1}{2}m\,\overline{(w^2)} = \tfrac{1}{2}Rt,$$

the average kinetic energy associated with each degree of freedom of a molecule is $\frac{1}{2}Rt$. When the size of the molecules is taken into account, the calculation of the pressure requires modification. This was first carried out by Van der Waals (Jeans, chap. VI).

136. *Atoms and molecules.*

The modern theory of the structure of the atom is in the first place due to J. J. Thomson. His researches on the conduction of electricity through gases[1] in 1897 disclosed the existence of negatively electrified particles, now called 'electrons,' whose mass is 1/1845 of the mass of an atom of hydrogen and whose charge (e) is 4.774×10^{-10} E.S.U.; the mass and charge of the electrons are independent of the nature of the gas used.

It thus became apparent that electrons might be an ultimate constituent of all atoms; Thomson[2,3] at first proposed a model of the atom in which electrons were in equilibrium in a uniformly diffused positive charge forming a minute sphere, and later gave a kinetic theory in which the electrons were revolving in orbits under the controlling influence of the positive charge. (The discovery of radioactivity had by now shown that within the atom there was a vast store of kinetic energy.)

At the same time as the electrons, there are produced in the gas tube positively electrified particles, now called 'positive rays,' of the same order of mass as the atom of the gas used.

Thomson[4] examined the positive rays from different substances by a deflexion method in which positive particles of a given kind produced a parabola on a photographic plate, whatever their speed; the study of these rays has since been greatly advanced by an experimental method of remarkable power due to Aston[5].

[1] *Conduction of Electricity through Gases*, 1903.
[2] *Electricity and Matter* (Silliman Lectures, 1903).
[3] *The Corpuscular Theory of Matter* (Royal Institution Lectures, 1906).
[4] *Rays of Positive Electricity*, 1913.   [5] *Isotopes*, F. W. Aston, 1923.

The researches of Rutherford and his pupils in which the instrument of the α-particle was used to disclose the nature of the atom have led to the atomic model proposed by Rutherford which is now generally used in theoretical work.

Rutherford's atom consists of a central positively charged nucleus around which revolve a number of negatively charged electrons. In a neutral atom the total charge of the electrons is equal and opposite to the charge of the nucleus. The number of electrons is equal to the 'atomic number' of the element. (This is a number indicating the position of the element in a table, in which, as Moseley proposed, the elements are arranged in the order of their atomic weights, beginning with hydrogen and allowing for obvious gaps for undiscovered elements suggested by the periodic law and X-ray spectra.)

On this theory the discharge through the gas tube in Thomson's experiments detaches electrons from the atom, leaving the nucleus and some or no electrons, which constitute a positive ray particle.

An atom of hydrogen, whose atomic number is 1, would consist of an electron (charge $-e$) revolving round a nucleus (charge $+e$). If the mass of the electron is the unit, that of the atom of hydrogen is 1845, so that the nucleus would effectively contain the mass of the atom. This nucleus, detached from its revolving electron, is a positive ray particle of hydrogen. This is called a 'proton.'

The positive nuclei of heavier atoms are built up of protons and electrons. For example if an element had an atomic weight a and atomic number n, the nucleus of its atom must contain a protons to give it the necessary mass, and there must be n electrons circulating round it. To make the charge on the nucleus equal and opposite to that of the electrons, the nucleus must also contain $(a - n)$ electrons, which would not affect its mass appreciably. Thus the nucleus would be an aggregate of a protons and

$(a - n)$ electrons of total charge ne and there would be n revolving electrons whose total charge is $-ne$.

The spectrum and the chemical properties of the element depend on its atomic number, *i.e.* the number of revolving electrons.

Thus two atoms, one with a nucleus (22 protons, 12 electrons) and 10 revolving electrons and the other with a nucleus of (20 protons, 10 electrons) and 10 revolving electrons, would have identical spectroscopic and chemical properties.

But the atomic weight of the former would be 22 and of the latter 20. Such atoms were called by Soddy 'isotopes.' The atomic weight of the gas 'neon' is 20·2, and Aston has shown that the gas is a mixture of two isotopes of atomic weights 20, 22.

Rutherford's concept of the atom has also been confirmed by the amazing exactness of Bohr's[1] calculations of the spectrum of hydrogen based on this model of the atom and the quantum mechanics; and also by the work of Sommerfeld[2] and others who have accounted for the fine structure of the lines (by using the relativity correction for the change of mass with velocity), and for the effects of electric and magnetic fields on the spectrum.

Molecules are aggregations of atoms; a molecule of mercury vapour is one atom; a molecule of oxygen or carbon monoxide is two atoms; a molecule of steam or carbonic acid is three atoms, and so on.

When a gas is heated, at first the atoms as a whole in the molecule are excited and heat rays of long wave-length (compared with light) are emitted. At very high temperatures, as in the production of arc and spark spectra, the emission of much shorter wave-lengths occurs also and is due to electron changes within the atom.

---

[1] *Phil. Mag.* vol. xxvi. 1913 and *Report on Radiation and the Quantum Theory*, J. H. Jeans, p. 35.
[2] A. Sommerfeld, *Atomic Structure and Spectral Lines* (English translation), 1923.

137. *The specific heat of a gas.*

We have seen that the average kinetic energy associated with each degree of freedom of a molecule is $\frac{1}{2}Rt$.

Consider a monatomic gas. If the temperature is within an ordinary range which is not sufficient to disturb the internal structure of the one atom composing the molecule, there are three degrees of freedom, *i.e.* those of translation of the molecule.

$$\therefore E = \tfrac{3}{2}Rt,$$

since the potential energy is negligible, the cohesion in gases being very small, as Joule's experiment shows.

Now for a perfect gas,

$$dE = c_v dt,$$

$$\therefore c_v = \frac{3R}{2}.$$

But $c_p - c_v = R.$ $\therefore c_p = \frac{5R}{2}.$

$$\therefore \gamma = \frac{c_p}{c_v} = \frac{5}{3} = 1.66.$$

This value agrees with that found for monatomic gases, such as helium and mercury vapour.

Consider a diatomic gas. The two atoms form a molecule with symmetry about the line joining the nuclei of the atoms. If the molecule is treated as rigid, so that there is no vibration of the nuclei relative to one another, there are only five effective degrees of freedom, *i.e.* three of translation and two of rotation, as rotation about the axis of symmetry could not be produced in encounters with other molecules. The energy E is therefore $\frac{5}{2}Rt$.

From this, $c_v = \dfrac{5R}{2}$, $c_p = \dfrac{7R}{2}$ and $\gamma = \dfrac{7}{5} = 1.4$.

For most diatomic gases, oxygen, hydrogen, carbon monoxide, this is found to be the case for moderate temperatures. But when these gases are strongly heated, the specific heat rises, so that γ falls. This is due to vibration of the nuclei and their electrons, which produces their heat and light radiation, of which the former is here the important factor.

Consider a triatomic gas, such as water vapour or carbonic acid. The three atoms form a molecule, which if rigid, has six degrees of freedom so that $E = 3Rt$, whence

$$c_v = 3R, c_p = 4R, \text{ and } \gamma = \tfrac{4}{3} = 1\cdot33.$$

The values of γ for water vapour and carbonic acid are rather less than this; for more complex molecules (which would if rigid still have no more than six degrees of freedom) the value of γ is much less. This is due to the energy of vibration within the molecule (so far neglected) becoming comparable with the energy of translation and rotation of the molecule as a rigid entity.

138. *Newtonian mechanics not sufficient.*

If the theorem of equipartition of energy, based upon the Newtonian scheme of mechanics, is applied to a molecule with six degrees of freedom, and also n degrees of vibratory freedom, then

$$E = 6\left(\tfrac{1}{2}Rt\right) + nRt = (n + 3)\,Rt.$$

$$\therefore \gamma = \frac{n + 4}{n + 3},$$

and is constant.

But it is known that γ falls with increasing temperature; so that equipartition alone will not suffice. Again the spectrum contains some thousands of lines indicating that n is some thousand, so that $\gamma \to 1$ on this theory, which is not the case.

Planck's formula.

Planck[1], starting with conceptions outside the Newtonian system of mechanics, found a formula

$$Rt\,\frac{x}{e^x - 1}, \text{ where } x = \frac{h\nu}{Rt},$$

[1] *La Théorie du Rayonnement et les Quanta*, a Report to the Solvay Congress at Brussels, 1911, p. 104. This is referred to later as *R. et Q.*

for the average energy (kinetic and potential) of a molecule whose frequency of vibration is ν. His theory, known as the 'quantum theory,' has as its foundation principle—'Energy of frequency ν can only be emitted in multiples of a quantum of energy ϵ, where $\epsilon = h\nu$, and h is a universal constant.' The constant h, Planck's constant, is $6 \cdot 55 \times 10^{-27}$; the frequency ν is the number of vibrations per second.

The graph of $y = \dfrac{x}{e^x - 1}$ is

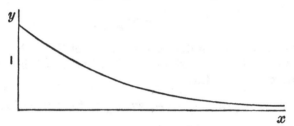

so that small values of x lead to the largest values in Planck's formula or low frequency vibrations are the most potent in effecting the average energy of a molecule. Thus when a gas is strongly heated the heat vibrations in the infra-red part of the spectrum are more important than the higher frequency vibrations in the visible part of the spectrum. The former are associated with the heavy nuclei of the atoms forming the molecule and the latter with the electrons revolving round the nuclei.

139. *Application to carbon monoxide.*

In the case of a diatomic molecule, such as that of carbon monoxide, we should expect one band in the infra-red corresponding to the period of vibration of the nuclei of the carbon and oxygen atoms relative to each other. Such a band of wave-length $\cdot00047$ cm. is known.

Since the wave-length $\times$ frequency $=$ velocity of the waves, $(\cdot00047)\,\nu = 3 \times 10^{10}$, the velocity of light (and any radiation).

$$\therefore \; \nu = \tfrac{3}{47} \times 10^{15} \text{ vibrations per second.}$$

$$x = \frac{h\nu}{Rt} = \frac{6\cdot55 \times 10^{-27} \times 3 \times 10^{15}}{47 \times 13\cdot8 \times 10^{-17} \times t} = \frac{3029}{t}.$$

At 0° C., $x = \dfrac{3029}{273\cdot1} = 11\cdot09.$

At 2000° C., the temperature of explosion of a gas mixture,

$$x = \frac{3029}{2273\cdot1} = 1\cdot33.$$

Now the energy associated with a molecule is

$$E = \tfrac{5}{2} Rt + Rt \frac{x}{e^x - 1},$$

the first term being due to the motion of the molecule as a whole and the second due to the vibration of the nuclei relative to each other.

$$\therefore \; E = \tfrac{5}{2} Rt + \frac{h\nu}{e^{\frac{h\nu}{Rt}} - 1}, \text{ writing in the value of } x.$$

$$\therefore \; \frac{dE}{dt} = \tfrac{5}{2} R + \frac{h\nu}{(e^x - 1)^2} \frac{h\nu}{Rt^2} e^x$$

$$= \left[\tfrac{5}{2} + \frac{x^2 e^x}{(e^x - 1)^2} \right] R.$$

$$\therefore \; \frac{c_v}{R} = \tfrac{5}{2} + \frac{x^2 e^x}{(e^x - 1)^2}. \qquad 1$$

Now $\dfrac{x^2 e^x}{(e^x - 1)^2}$ when $x = 11\cdot09$ is of order $\dfrac{x^2}{e^x}$ or $\dfrac{(11\cdot09)^2}{e^{11\cdot09}}$ and is negligible, and when $x = 1\cdot33$, its value is $\cdot87$.

$$\therefore \; \frac{c_v}{R} \text{ rises from } 2\cdot5 \text{ at } 0° \text{ C. to } 3\cdot37 \text{ at } 2000° \text{ C.}$$

$$\therefore \; \frac{c_p}{R} \text{ rises from } 3\cdot5 \text{ at } 0° \text{ C. to } 4\cdot37 \text{ at } 2000° \text{ C.}$$

$$\therefore \; \gamma \text{ falls from } \tfrac{7}{5} \text{ to } \frac{4\cdot37}{3\cdot37} \text{ in that range,}$$

or γ falls from $1\cdot4$ to $1\cdot3$ as the temperature rises from 0° to 2000° C.

[1] Planck, R. et Q. p. 112.

Thus Planck's quantum theory does account for the rise of the specific heat of a gas with the temperature, whereas the Newtonian theory requires it to be constant.

140. *Carbonic acid.*

This triatomic gas may have three degrees of freedom of vibration of the two oxygen nuclei and the carbon nucleus. There are three bands in the infra-red of wave-lengths 2·7, 4·3, 14·7 times 10^{-4} cm.

Even at 0° C., the wave-length $14·7 \times 10^{-4}$ produces an appreciable effect in Planck's formula and reduces γ from 1·33 to 1·30.

At very high temperatures the expression $Rt \dfrac{x}{e^x - 1}$ may be nearly Rt for each frequency (for if t is large, $x \to 0$), and thus a term nearly $3Rt$ would be added to the $3Rt$ there would be without this vibrational energy, or the total is nearly $6Rt$. Therefore c_v may be nearly $6R$ at very high temperatures, and c_p would be $7R$, so that γ would fall to $\frac{7}{6}$ as a limit, or 1·17, the normal value being 1·33 for such a gas.

This radiant energy has been actually measured by Hopkinson[1] in gas explosions.

141. *Halogen vapours.*

For the heavy atoms of chlorine, bromine, iodine, which form diatomic molecules, the vibrations are so slow that even at ordinary temperatures the effect of vibration on the specific heats is appreciable and there are abnormally high specific heats.

When for instance hydrogen is substituted for iodine in a diatomic molecule of iodine so as to produce hydriodic acid, the much lighter molecule, still diatomic, vibrates so rapidly that its vibration contributions to the energy are small and the specific heat has a normal value.

[1] *Proc. Roy. Soc.* 1910.

142. *Application to gas-engine theory.*

The rise of specific heat with temperature is of importance in the theory of a gas-engine, for it means that the rise of temperature resulting from the explosion of a given gas mixture is not so great as it would be if the specific heat remained constant.

It has been shown that the rise of c_v or c_p due to a molecular vibration of frequency ν, produced by a high temperature is $\dfrac{Rx^2e^x}{(e^x - 1)^2}$, where $x = \dfrac{h\nu}{Rt}$.

Take a definite instance, that of carbon monoxide. It has been seen that $x = \dfrac{3029}{t}$. If $R\eta$ denote the rise of the specific heat, $\eta = \dfrac{x^2e^x}{(e^x - 1)^2}$; writing $x = \dfrac{1}{\xi}$, $\eta = \dfrac{\dfrac{1}{\xi^2}e^{\frac{1}{\xi}}}{(e^{\frac{1}{\xi}} - 1)^2}$,

where $\xi = \dfrac{t}{3029}$.

The curve given by this ξ-η equation is a curve showing the relation between the rise of specific heat and the temperature; ξ is the temperature divided by 3029 and η is the rise of specific heat divided by R. The curve is of the form

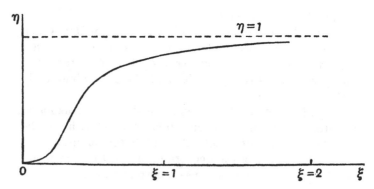

For a range of temperature 0° to 2000° C., ξ ranges from

$\dfrac{273 \cdot 1}{3029}$ to $\dfrac{2273 \cdot 1}{3029}$ or $\cdot 09$ to $\cdot 75$. The figure shows that for this range, the curve is approximately parabolic.

143. *Practical formulae for the variation of the specific heat.*
A formula $c_v = a + bt + ct^2$ has accordingly been used (a parabolic curve). Hence $c_p = a' + bt + ct^2$, where
$$a' - a = R.$$

The adiabatic curves for these high temperatures for which c_v, c_p vary are given by
$$c_v dt + p dv = 0$$

$$(a + bt + ct^2) \frac{dt}{t} + R \frac{dv}{v} = 0$$

or $a \log t + bt + \dfrac{ct^2}{2} + R \log v = \text{constant.}$

Thus a v-t curve can be drawn.

Using $pv = Rt$, a p-v curve can be deduced.

144. *Specific heat of solids at low temperatures.*

Dulong and Petit showed that the product of the specific heat and the atomic weight, for a large number of solid elements, has at ordinary temperatures an approximately constant value. This product, called the '*atomic heat*,' is about 6·2.

If m is the mass of an atom and n the number per gram, $mn = 1$. If the number of degrees of freedom of the atom[1] is s, the average kinetic energy associated with each degree of freedom is $\frac{1}{2}Rt$; and as in a vibration the mean energy is half kinetic and half potential, the total energy for each degree of freedom is Rt.

Therefore the total energy E per gram $= nsRt$ ergs.

$$c_v = \frac{dE}{dt} = nsR.$$

Let a be the atomic weight of the element assuming that of oxygen to be 16.

---

[1] *R. et Q.* p. 63.

If m_0 is the mass of an atom of oxygen,

$$\frac{a}{16} = \frac{m}{m_0}. \qquad \therefore n = \frac{1}{m} = \frac{16}{am_0}.$$

$$\therefore c_v = \frac{16sR}{am_0}.$$

Now $R = 13\cdot8 \times 10^{-17}$, and $m_0 \times$ (number of atoms of oxygen per c.c.) = mass of 1 c.c. of oxygen = $\cdot001429$ gr.

$$\therefore m_0 \, (2 \times 2\cdot7 \times 10^{19}) = \cdot001429.$$

[Avogadro's number $2\cdot7 \times 10^{19}$ molecules per c.c. is multiplied by 2, since the molecule of oxygen contains two atoms.]

$$\therefore \frac{R}{m_0} = \frac{13\cdot8 \times 10^{-17} \times 5\cdot4 \times 10^{19}}{\cdot001429} = 521 \times 10^4.$$

$$\therefore ac_v = 16 \times 521 \times 10^4 \times s \text{ in work units}$$

$$= \frac{16 \times 521 \times 10^4 \times s}{41\cdot8 \times 10^6} \text{ in heat units,}$$

or the atomic heat, $A = (1\cdot994) \, s$.

If $s = 3$, then $A = 5\cdot98$, approximately the observed value. On this theory, based on equipartition (Newtonian theory) the atomic heat is $(1\cdot994) \, s$, where s is a constant, so that no variation of specific heat with temperature is possible.

The energy is $3nRt$ per gram and $c_v = 3nR$, where n is the number of molecules per gram.

145. *Fall of atomic heat at low temperatures.*

In 1911 Nernst[1] determined a large number of specific heats at constant pressure through a wide range of temperatures, and corrected them to specific heats at constant volume (§ 75). It was found that the atomic heat was constant at ordinary temperatures and about 6, but that there was a marked diminution at very low temperatures. The curves obtained were of the type shown in the figure opposite.

[1] *Zeits. für Elektrochem.* 1911.

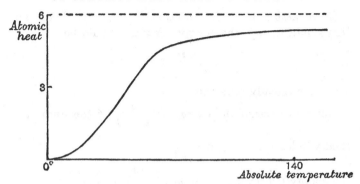

Now Planck's formula $Rt \dfrac{x}{e^x - 1}$ agrees with the New-tonian one for small values of x; the differences between quantum theory and Newtonian theory arise when x is not small.

Since $x = \dfrac{h\nu}{Rt}$, we should expect differences for large values of ν or small values of t, $i.e.$ high frequencies or low temperatures. It is to the quantum theory therefore that we must look to account for these anomalies at very low temperatures.

146. *Debye's theory.*

In 1912, Debye[1] gave a theory of the variation of atomic heat which leads to curves agreeing almost exactly with those observed by Nernst. The atoms of a solid do not possess independent free vibration; the oscillations of one are affected by all the others. Assuming that there is an upper limit ν_m to the values of the frequencies possible, Debye showed that the number of vibrations with frequencies between ν and $\nu + d\nu$ must be

$$\frac{9n\nu^2 d\nu}{\nu_m^3},$$

where n is the number of atoms per gram.

[1] *Ann. der Physik*, 1912.

If we assign to each vibration the energy Rt of the Newtonian theory, the energy per gram would be

$$\int_0^{\nu_m} \frac{9n\nu^2 d\nu}{\nu_m{}^3} Rt = 3nRt,$$

the value already obtained.

But if we assign the energy $Rt \dfrac{x}{e^x - 1}$ of the quantum

theory, where $\qquad\qquad x = \dfrac{h\nu}{Rt},$

the energy per gram, $\quad E = 9n \int_0^{\nu_m} \dfrac{\nu^2 d\nu}{\nu_m{}^3} \dfrac{h\nu}{e^{\frac{h\nu}{Rt}} - 1}.$

$$\therefore \; c_v = \frac{dE}{dt} = \int_0^{\nu_m} \frac{9nh^2 \nu^4 e^{\frac{h\nu}{Rt}} d\nu}{Rt^2 \nu_m{}^3 (e^{\frac{h\nu}{Rt}} - 1)^2}.$$

Writing $\qquad\qquad x = \dfrac{h\nu}{Rt}$ and $x_m = \dfrac{h\nu_m}{Rt},$

$$c_v = \frac{9nR}{x_m{}^3} \int_0^{x_m} \frac{x^4 e^x dx}{(e^x - 1)^2}.$$

At ordinary temperatures x_m is small, so that

$$\int_0^{x_m} \frac{x^4 e^x dx}{(e^x - 1)^2} \to \int_0^{x_m} \frac{x^4 dx}{x^2} \to \frac{x_m{}^3}{3},$$

so that $c_v \to 3nR$, the Newtonian value.

Denoting this by c_{v0}, we have

$$\frac{c_v}{c_{v0}} = \frac{3}{x_m{}^3} \int_0^{x_m} \frac{x^4 e^x dx}{(e^x - 1)^2},$$

a definite function of x_m; let this be $f(x_m)$.

This too is $\dfrac{A}{A_0}$ the ratio of the atomic heat A to the constant value A_0 observed at ordinary temperatures.

Hence $\qquad \dfrac{A}{A_0} = f(x_m) = f\left(\dfrac{h\nu_m}{Rt}\right) = F\left(\dfrac{\nu_m}{t}\right),$

the function F being the same for all elements. The curve F obtained from the integral was found to have the form of the Nernst experimental curve and could be fitted to it by adjustment of ν_m; the values of ν_m required for the different substances were found to be in good agreement with the values calculated from their elastic constants.

147. *Atomic heat near the absolute zero.*

If the temperature is near the absolute zero, x_m is large and

$$\int_0^{x_m} \frac{x^4 e^x dx}{(e^x - 1)^2} \to \int_0^\infty \frac{x^4 e^x dx}{(e^x - 1)^2}$$

or

$$\int_0^\infty x^4 \left(e^{-x} + 2e^{-2x} + 3e^{-3x} + \ldots\right) dx.$$

Since

$$\int_0^\infty x^4 e^{-cx} dx = \frac{24}{c^5},$$

the integral

$$\to 24\left(\frac{1}{1^5} + \frac{2}{2^5} + \frac{3}{3^5} + \ldots\right)$$

$$= 24\left(\frac{1}{1^4} + \frac{1}{2^4} + \frac{1}{3^4} + \ldots\right)$$

$$= \frac{4\pi^4}{15}$$

$$= 25\cdot97.$$

$$\therefore \quad \frac{A}{A_0} \to \frac{3}{x_m{}^3} \times 25\cdot97$$

or

$$\to \frac{77\cdot9}{x_m{}^3}, \quad i.e.\ 77\cdot9\left(\frac{Rt}{h\nu_m}\right)^3 \quad \ldots\ldots\ldots\ldots(1).$$

Thus for a given substance, A varies as t^3 near the absolute zero. This agrees with the observations of Kamerlingh Onnes and Keesom (1915)[1] at temperatures as low as 15° absolute; the values of ν_m found from their results by the equation (1) agree closely with the values calculated from the elastic constants.

---

[1] *Commun. Phys. Lab. of Leiden,* 147 a.

CHAPTER XIV

RADIATION

148. *Radiation.*

Electromagnetic waves are known of a wide range of wave-length; electric waves, infra-red rays, light, ultra-violet rays, X-rays and γ-rays; a sequence of wave-lengths diminishing from many hundreds of metres for electric waves down to lengths of order 10^{-10} cm. for γ-rays. These are all included in the term 'radiation.'

A train of such waves has a definite amount of energy associated with each wave-length, so that as the waves travel (with the velocity of 3×10^{10} cm. per second) a stream of energy travelling at that rate is drawn out from the body producing the radiation.

If a body is kept at a given temperature and steadily emits radiation without undergoing any change of structure, the radiation is called 'pure temperature radiation.' Lampblack is a substance which can emit or absorb almost all wave-lengths of temperature radiation; an ideal body which can emit all wave-lengths is called a 'black body.'

If a black body is heated it gives out radiations of all wave-lengths, but at first, only the long infra-red waves (radiant heat) have sufficient intensity to be perceived by the use of a thermometer or bolometer. As the temperature rises, the intensity of the energy corresponding to all wave-lengths increases and at 525° C. that of the shorter waves of light is sufficient for them to affect the eye. The light is at first of long wave-length (red) so that the body is 'red hot' and then smaller and smaller wave-lengths become perceptible, until at about 1200° C. the body emits visible light of all wave-lengths and is 'white hot'; at higher temperatures the ultra-violet rays become apparent, until at 2325° C. full temperature radiation is obtained (Lummer and Pringsheim)[1].

[1] *Berichte der Deuts. Ges.* 1903.

149. *The pressure of radiation.*

Maxwell[1] in 1873 deduced from the electromagnetic theory that radiation incident normally on a surface at rest produces a pressure equal to the energy per unit volume of the radiation. The theory of radiation pressure has been developed by the researches of Larmor[2], who found that for radiation incident normally on a surface moving in the same direction with velocity u, the radiation pressure is

$$\frac{c^2 - u^2}{c^2 + u^2} \text{ of Maxwell's value for a fixed surface,}$$

c being the velocity of light. This agrees with Maxwell's value when u/c is small.

150. If the radiation is incident obliquely at an angle θ with the normal, the pressure across $BC = E$, where E is the energy per unit volume; or the force $= E\,(BC) = E \cos \theta$, and resolved normally to BC, this yields $E \cos^2 \theta$ as the pressure on AB.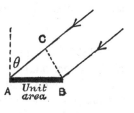

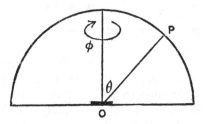

151. Now consider radiation falling equally in all directions upon a unit area at O, on one side of it.

The pressure is the mean value of $E \cos^2 \theta$, taken for all directions from O to the hemisphere shown.

[1] *Electricity and Magnetism*, vol. II.
[2] *On the dynamics of radiation*, Intern. Congress of Math. 1912, vol. I. p. 197.

If $d\omega$ is the solid angle of a small cone whose axis is OP, the mean value is

$$\frac{\iint E \cos^2 \theta \, d\omega}{\iint d\omega} = \frac{\int_0^{\pi/2} \int_0^{2\pi} E \cos^2 \theta \sin \theta \, d\theta \, d\phi}{2\pi}$$

$$= E \int_0^{\pi/2} \cos^2 \theta \sin \theta \, d\theta = \tfrac{1}{3} E.$$

152. Radiation and temperature.

Stefan's law. In 1817 Dulong and Petit[1], by observing the rate of cooling of thermometers at different temperatures in a copper globe surrounded by water, deduced from the cooling curves a law connecting the emission of radiant energy per second (E) with the temperature (t) of the thermometer. The law was expressed by

$$E = a \, (1 \cdot 0077)^t + b.$$

In 1879 Stefan[2] proposed an entirely new formula that E is proportional to t^4. He was led to this by a result of Tyndall's, who had found that the radiation from a platinum wire at 1200° C. was 11·7 times that at 525° C.

Now $\left(\dfrac{1200 + 273}{525 + 273}\right)^4$ is 11·6, so that in this instance the radiation varied as the fourth power of the absolute temperature. He examined the work of Dulong and Petit and found that if their results were corrected for conduction due to the surrounding medium, they were more in agreement with his formula than with their own.

In 1875 Bartoli had applied thermodynamical principles to radiation; but in 1884 Boltzmann[3] regarding full radiation in an enclosure as exerting pressure on its boundary like the working substance in an engine, took it through a Carnot cycle, and using Maxwell's formula for the radiation pressure, deduced Stefan's law.

---

[1] *Ann. de Chimie*, VII. 1817.
[2] *Wien. Akad. Ber.* LXXIX. 1879.
[3] *Wied. Ann.* XXI. 1884.

153. *Boltzmann's deduction of Stefan's law.*

Consider a cylinder fitted with a piston, the sides of the cylinder and the piston being perfect reflectors, *i.e.* non-permeable to radiation. Let the end of the cylinder be perfectly absorbent. Let there be two sources of radiation at temperatures t_1, t_2 ($t_1 > t_2$), and also a slab of material impervious to radiation.

Consider the cylinder with the piston pushed home to the end of the cylinder and suppose it at temperature t_1.

(i) Place the end in contact with the hotter source (t_1) and allow the piston to rise; radiation enters the cylinder and the radiation pressure does work.

(ii) Place the slab over the end of the cylinder so that now no radiation can escape; suppose the piston to move on, so that the system expands adiabatically and the temperature falls to t_2.

(iii) Remove the slab and place the end in contact with the colder source (t_2) and let the piston be driven home so that all the radiation is expelled.

(iv) Place the end of the cylinder in contact with the hotter source. No appreciable radiation is used up in bringing it up to temperature (t_1), and the cycle is complete.

Let the energy of radiation per unit volume be E at temperature t and radiation pressure p.

Let v_1, v_2 be the volumes of the space at the beginning and end of the adiabatic process (ii). (We shall assume, what is demonstrable, that the fullness of the radiation is not affected by adiabatic change.)

In process (i) the internal energy increases by $E_1 v_1$ and the work done in expansion is $p_1 v_1$, so that the heat taken in from the source at temperature t_1 is $(E + p_1) v_1$. So in process (ii) the heat given out at temperature t_2 is $(E_2 + p_2) v_2$.

By the theory of Carnot's cycle,

$$\frac{(E_1 + p_1) v_1}{t_1} = \frac{(E_2 + p_2) v_2}{t_2}.$$

In the adiabatic process (ii), the increase of energy + the work done by the system = the heat taken in = 0.

$$\therefore\ E_2 v_2 - E_1 v_1 + \int_{v_1}^{v_2} p\,dv = 0.$$

For an infinitesimal change for which $v_1 = v$, $v_2 = v + dv$, these equations become

$$\left. \begin{array}{r} d\left[(E + p)\dfrac{v}{t}\right] = 0 \\ d\,(Ev) + p\,dv = 0 \end{array} \right\}.$$

Therefore

$$\left. \begin{array}{r} \{(dE + dp)\,v + (E + p)\,dv\}\,t - (E + p)\,v\,dt = 0 \\ Edv + vdE + pdv = 0 \end{array} \right\}.$$

$$\therefore\ dp\,.\,t = (E + p)\,dt.$$

Now $p = \dfrac{E}{3}$, from Maxwell's theory (§ 151);

$$\therefore\ \frac{dE}{E} = \frac{4dt}{t}.$$

$$\therefore\ E = Ct^4, \text{ where } C \text{ is a constant.}$$

This is Stefan's experimental law.

154. *Deduction by Thomson's equation.*
The Thomson equation is

$$\psi = \epsilon + t\left(\frac{\partial\psi}{\partial t}\right)_v,$$

where ϵ is the total internal energy. Consider a volume v of radiation and let it increase to $v + dv$ at constant temperature.

The increase in ψ is equal to the work done on the system $= -pdv$. The increase of ϵ is Edv, since E is constant when t is constant.

Therefore by Thomson's equation

$$-pdv = Edv - t\left(\frac{\partial p}{\partial t}\right)_v dv.$$

$$\therefore\ -p = E - t\left(\frac{\partial p}{\partial t}\right)_v.$$

And $p = \dfrac{E}{3}$. $\therefore 4E = t\left(\dfrac{\partial E}{\partial t}\right)_v$.

$\therefore E = Ct^4$, at constant volume.

155. *Experimental verification of Stefan's law.*

Consider a hollow metal vessel with a small hole in it. Any radiation entering the hole from outside would be partly absorbed by the walls and partly reflected. The chance of the reflected radiation ever emerging from the hole is very small indeed, and such minute portion as did so would have undergone so many reflections that its intensity would be very small. It may therefore be said that a small hole in a hollow body is a perfect absorber of heat, that is, it is the experimental realisation of the perfect 'black' body.

Hence if a small hole be made in the wall of a hollow metal vessel heated to some given temperature, the radiation from the hole will be that of the 'ideal' black body at that temperature.

Lummer and Pringsheim[1] found by bolometric measurements between 100° C. and 1300° C. that the radiation from a small hole in a hollow shell or 'black body' radiation followed Stefan's law.

156. *The spectrum of a black body.*

By the use of a prism of fluorspar, 'black body' radiation can be dispersed into a spectrum. If the energy of the part of the spectrum corresponding to wave-lengths between λ and $\lambda + d\lambda$, where $d\lambda$ is a small definite range of wave-length, is measured by a bolometer and is $E_\lambda \cdot d\lambda$, E_λ can be plotted as ordinate with λ as abscissa for all the wave-lengths of the spectrum. The distribution of energy amongst the different wave-lengths is thus known by experiment. It is found that there is a wave-length (λ_m) for which E_λ is a maximum at any given temperature. Wien[2], in 1893, by theoretical considerations,

[1] *Wied. Ann.* LXIII. 1897.
[2] *Sitzungsberichte Berl. Akad.* 1893.

showed that for different temperatures, λ_m is proportional to $\dfrac{1}{t}$ and that $(E_\lambda)_m$ is proportional to t^5 (§ 158).

157. *The effect of change of temperature on the spectrum of a black body.*

(α) *Effect of adiabatic change on full radiation.*

Consider an enclosure full of radiation with perfectly reflecting boundaries and let its volume increase from v to $v + dv$.

The process is adiabatic, so that the work done is equal to the decrease of internal energy. Therefore

$$pdv = - d\,(Ev).$$

Now
$$p = \frac{E}{3}.$$

$$\therefore\ Edv + 3d\,(Ev) = 0.$$

$$\therefore\ 4Edv + 3vdE = 0.$$

$$\therefore\ v^4 E^3 = \text{constant.}$$

By Stefan's law, $E \propto t^4$,

$$\therefore\ vt^3 = \text{constant.}$$

If the enclosure is *spherical*, of radius r, then it follows (since v varies as r^3) that Er^4 is constant, and rt is constant, during adiabatic change.

(β) *The Döppler effect.*

This is an effect on the wave-length due to the motion of the boundary of the enclosure, analogous to the well-known Döppler effect in the theory of sound waves.

Consider a train of waves of length λ reflected normally from a perfectly reflecting surface moving with velocity u in the same direction as the waves whose velocity is c, where u/c is supposed small.

Consider the disturbance at A at time $t = 0$; let the reflecting plane be then distant a from A. The disturbance reaches the plane at time $t = t_1$ when the distance of the

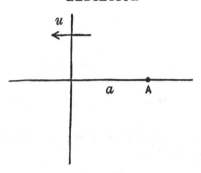

plane from A has become a_1; it is reflected and returns to A at time t'. Then

$$\left.\begin{array}{l} a_1 - a = vt_1 \\ a_1 = ct_1 \\ 2a_1 = ct' \end{array}\right\}.$$

$$\therefore \ a_1 - a = \frac{va_1}{c}.$$

$$a_1 = \frac{ac}{c-v}.$$

$$\therefore \ t' = \frac{2a}{c-v}.$$

Thus a disturbance leaving A at time $t = 0$, returns at time $t = \dfrac{2a}{c-v}$.

Let the disturbance at A be periodic (as in a wave) of period τ. Then at time $t = \tau$ a disturbance equal to the one just considered starts from A and the distance a has now become $a + v\tau$. This disturbance therefore returns *after* time $\dfrac{2(a+v\tau)}{c-v}$ and therefore *at* time $\dfrac{2(a+v\tau)}{c-v} + \tau$.

The interval between these two times of return to A must be the period τ' of the reflected wave.

$$\therefore \ \tau' = \frac{2(a+v\tau)}{c-v} + \tau - \frac{2a}{c-v}$$

or $$\tau' = \frac{c + v}{c - v}\,\tau.$$

$$\therefore \tau' = \left(1 + \frac{2v}{c}\right)\tau, \text{ approximately}$$

or $$d\tau = \frac{2v}{c}\,\tau, \text{ writing } \tau' = \tau + d\tau.$$

Now for any wave, $c\tau = \lambda$ and c is constant for all wave-lengths, therefore

$$\frac{d\tau}{\tau} = \frac{d\lambda}{\lambda}.$$

Hence $$d\lambda = \frac{2v}{c}\,\lambda.$$

For oblique incidence at an angle θ with the normal to the reflecting surface, the effective part of v is that perpendicular to the wave front or $v\cos\theta$, and

$$d\lambda = \frac{2v \cos\theta}{c}\,\lambda.$$

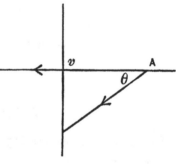

Now consider a spherical enclosure containing full radiation and suppose the radius (r) of its boundary is changing.

If θ is the angle of incidence of radiation of wave-length λ, the length of the path between successive reflections is $2r \cos\theta$. The number of reflections per second is therefore

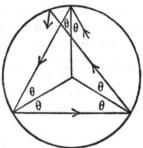

$$\frac{c}{2r \cos\theta}.$$

At each one $\quad d\lambda = \dfrac{2v \cos \theta}{c} \lambda.$

Therefore the increase of λ per second is

$$\left(\frac{2v \cos \theta}{c} \lambda\right)\left(\frac{c}{2r \cos \theta}\right)$$

$$= \frac{v\lambda}{r}.$$

$$\therefore \frac{\dot{\lambda}}{\lambda} = \frac{v}{r} = \frac{\dot{r}}{r}.$$

Therefore λ/r is constant, as the radius of the boundary changes.

(γ) *Wien's formula,* $E_\lambda = t^5 f(\lambda t).$

Consider a spherical enclosure of full radiation. Using the notation of § 156 and the results proved in (α), (β) we have that when r becomes r', λ becomes λ' and t becomes t',

$$\left.\begin{array}{c} E_\lambda \, d\lambda \, r^4 = E_{\lambda'} \, d\lambda' \, r'^4 \\ rt = r't' \\ \dfrac{\lambda}{r} = \dfrac{\lambda'}{r'} \end{array}\right\}.$$

For a wave $\lambda + d\lambda,$ $\quad \dfrac{\lambda + d\lambda}{r} = \dfrac{\lambda' + d\lambda'}{r'}.$

$$\therefore \frac{d\lambda}{r} = \frac{d\lambda'}{r'}.$$

Hence $\quad E_\lambda r^5 = E_{\lambda'} r'^5 \, ;$

$$\left.\begin{array}{c} \therefore \dfrac{E_\lambda}{t^5} = \dfrac{E_{\lambda'}}{t'^5} \\ \lambda t = \lambda' t' \end{array}\right\} \dots\dots\dots\dots(1).$$

and

Now plot two curves, the first having

$$y = \frac{E_\lambda}{t^5}, \ x = \lambda t$$

and the second $\quad y = \dfrac{E_{\lambda'}}{t'^5}, \ x = \lambda' t'.$

Then on account of equations (1), the two curves are identical for any values of t and t'. Let the curve be $y = f(x)$.

Then $\dfrac{E_\lambda}{t^5} = f(\lambda t)$ for all temperatures, or

$$E_\lambda = t^5 f(\lambda t).$$

This is Wien's formula for the form of E_λ.

It may also be written

$$E_\lambda = \frac{(\lambda t)^5 f(\lambda t)}{\lambda^5}$$

or

$$E_\lambda = \frac{1}{\lambda^5} F(\lambda t).$$

158. *Wien's displacement law.*

We have seen that if $y = \dfrac{E_\lambda}{t^5}$, $x = \lambda t$, then $y = f(x)$, where the function f is the same for all temperatures.

Then if $Y = y t^5$, $X = \dfrac{x}{t}$, the curve between Y and X is that between E_λ and λ, at temperature t.

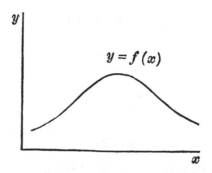

It is only necessary to draw the curve $y = f(x)$ and increase the scale of the ordinate in the ratio $t^5 : 1$ and that of the abscissa in the ratio $1 : t$ to obtain the E_λ, λ curve for any temperature t.

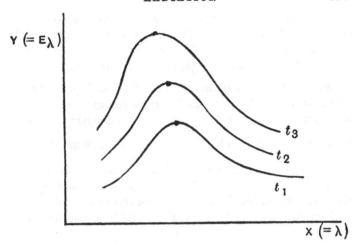

If the curve $y = f(x)$ has a maximum at (x_0, y_0), each of the E_λ, λ curves has a maximum given by

$$Y_m = y_0 t^5, \quad X_m = \frac{x_0}{t},$$

where x_0, y_0 are independent of the temperature, and depend only on the form of the function f.

Therefore $(E_\lambda)_m$ varies as t^5 and the corresponding λ_m varies as $\frac{1}{t}$.

Thus the maximum intensity corresponds to a shorter and shorter wave-length as the temperature rises. This is Wien's 'displacement' law.

159. *Experimental verification of Wien's law.*

The laws $\lambda_m \propto \frac{1}{t}$ and $(E_\lambda)_m \propto t^5$ have both been verified by Lummer and Pringsheim[1] by examining the radiation from a given narrow portion of the black body

[1] *Deutsch. Phys. Ges.* 1899.

158   RADIATION

spectrum for different wave-lengths; the temperatures used varied from 620° to 1646° (Thomson degrees). Curves of the types shown on p. 157 were found connecting E_λ with λ. It was found that $(\lambda_m)\,t = \cdot294$ cm. $\times$ deg.

160. *Distribution of energy in the black body spectrum.*

We have seen how general thermodynamical principles lead to the result that the energy of wave-lengths between λ and $\lambda + d\lambda$ is $E_\lambda d\lambda$ where $E_\lambda = \dfrac{1}{\lambda^5} F\,(\lambda t)$. The graph of F is known from experiment. To find theoretically the actual form of F as a function, it is necessary to use some hypothesis as to the molecular mechanism of emission of radiation.

Using Newtonian dynamics, Rayleigh[1] found the formula

$$E_\lambda = \frac{8\pi R t}{\lambda^4}$$

for large wave-lengths, which was shown by Jeans[2] to hold for any wave-length. But experiment shows that for low temperatures and short wave-lengths the intensity E_λ is incomparably smaller than that given by the above formula; also the total energy $\displaystyle\int_0^\infty E_\lambda d\lambda$ would be infinite for any finite value of t.

Other formulae have been obtained by Wien

$$\left(E_\lambda = \frac{a}{\lambda^5} e^{-\frac{b}{\lambda t}}\right),$$

and by J. J. Thomson[3]

$$\left(E_\lambda = \frac{8\pi R t}{\lambda^4} e^{-\frac{c}{\lambda t}}\right).$$

These formulae give a small E_λ for a small λ, but do not agree sufficiently well with experiments over a wide range of λ.

161. *Planck's formula.*

The formula of Planck[4] agrees closely with experiment

[1] *Phil. Mag.* 1900.   [2] *Phil. Mag.* 1909.
[3] *Phil. Mag.* 1907.   [4] *Annalen der Physik*, 1901, *R. et Q.* p. 93.

and is based on an assumption which is the foundation of the quantum theory. His formula is

$$E_\lambda = \frac{8\pi Rt}{\lambda^4}\left(\frac{\theta}{e^\theta - 1}\right), \text{ where } \theta = \frac{h\nu}{Rt}.$$

This is the Rayleigh-Jeans formula multiplied by the factor $\dfrac{\theta}{e^\theta - 1}$, where ν is the frequency corresponding to wave-length λ and h is Planck's constant such that $h\nu$ is the quantum of energy.

Since the wave-length = period × velocity of radiation,

$$\lambda = \left(\frac{1}{\nu}\right)c, \text{ where } c = 3 \times 10^{10} \text{ cm. per sec. or } \nu = \frac{c}{\lambda}.$$

Hence
$$E_\lambda = \frac{8\pi h\nu}{\lambda^4\left(e^{\frac{h\nu}{Rt}} - 1\right)} = \frac{8\pi hc}{\lambda^5\left(e^{\frac{hc}{R\lambda t}} - 1\right)}.$$

This is of the form $\dfrac{1}{\lambda^5}F(\lambda t)$ required by Wien's law.

Thus if $y = E_\lambda$ and $x = \lambda$, this formula is

$$y = \frac{8\pi hc}{x^5\left(e^{\frac{a}{x}} - 1\right)}, \text{ where } a = \frac{hc}{Rt}.$$

The maximum of y for a given temperature is given by the minimum of $x^5(e^{\frac{a}{x}} - 1)$, or by

$$5x^4(e^{\frac{a}{x}} - 1) - x^5 e^{\frac{a}{x}}\left(\frac{a}{x^2}\right) = 0,$$

or
$$e^{\frac{a}{x}} - 1 = \frac{a}{5x}e^{\frac{a}{x}}$$

or
$$1 - e^{-\frac{a}{x}} = \frac{1}{5}\frac{a}{x}$$

or
$$1 - e^{-\theta} = \frac{1}{5}\theta, \text{ where } \theta = \frac{a}{x} = \frac{hc}{Rt\lambda} = \frac{h\nu}{Rt}.$$

The root of this equation for θ is 4·96 approximately, so that the maximum of y or of E_λ is given by $\dfrac{hc}{Rt\lambda} = 4\cdot96$;

or $(E_\lambda)_m$ corresponds to $\lambda_m = \dfrac{hc}{(4 \cdot 96)\,Rt}$ at any given temperature t.

The form of the curve

$$y = \dfrac{C}{x^5 \left(e^{\frac{a}{x}} - 1\right)},$$

where C is the constant $8\pi hc$, is of the type shown on p. 157 and agrees closely with experiment for a large range of wave-lengths.

Lummer and Pringsheim's experiments showed that

$$(\lambda_m)\,t = \cdot 294 \text{ cm.} \times \text{deg.}$$

Planck's formula gives

$$(\lambda_m)\,t = \dfrac{hc}{(4 \cdot 96)\,R}, \text{ above.}$$

$$\therefore \quad \dfrac{hc}{(4 \cdot 96)\,R} = \cdot 294$$

or
$$h = \dfrac{\cdot 294 \times 4 \cdot 96\,R}{c}$$

$$= \dfrac{\cdot 294 \times 4 \cdot 96 \times 13 \cdot 8 \times 10^{-17}}{3 \times 10^{10}}$$

$$= 6 \cdot 7 \times 10^{-27}.$$

This constant h of Planck has been found in very diverse ways from observations of spectra, the photo-electric effect, and specific heats at low temperatures. The most probable value of the constant h is $6 \cdot 55 \times 10^{-27}$ ergs sec.

162. *Stellar spectra.*

The spectra of the stars are of types indicated in order by the letters $OBAFGKMNR$ (in the figure of p. 161), the types merging into one another; they correspond to successive stages in the evolution of a star. The sequence from R to O is the spectra of stars whose colours range from red to blue and whose temperatures range from 2000° C. for an R type of star to 20,000° C. for an O type.

This increase of temperature is accompanied by increased ionisation, which becomes so great that for the Wolf-Rayet stars, of O type, only the lines of hydrogen and helium are present in the spectrum.

The effective temperature of a star is the temperature of a black body whose distribution of energy in the spectrum is the same as that of the star. The position of λ_m in the spectrum of the star determines its temperature, by Wien's Law that λ_m varies as $1/t$ (§ 158). H. N. Russell pointed out that if the absolute magnitudes of the stars are plotted as ordinates with spectral types as abscissae, the stars lie in the main along two straight bands, shown in the figure by the dotted lines.

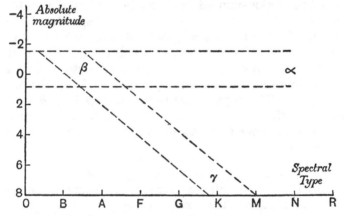

Thus the stars form two series, one very bright and of magnitude independent of spectral type (in the band $\beta\alpha$) and the other diminishing in brightness as the spectral type changes from O to R (in the band $\beta\gamma$): the former were designated by Russell 'giant' stars, and the latter 'dwarf' stars.

Observations of binary stars show that the densities of the stars steadily increase along the bands from α to β to γ. For stars of types A, B, the density increases from ·05 to ·5, while for stars of type G the density is either of order 1 or ·0001 (the density of water being the unit), the higher value being that of the dwarf (in $\beta\gamma$) and the lower that of the giant (in $\beta\alpha$).

The sun is a star of type G of density 1·4 (a dwarf), surface temperature of order 6000° C and of the fifth magnitude (absolute).

163. *Stellar evolution.*

The Lane-Ritter theory supposes that a star is at first a highly rarefied gas which contracts under its own gravitation, the potential energy thus lost being converted into heat. The temperature rises and the spectrum advances from R to O, the radiation of energy from the star being less than what is supplied by contraction and other possible causes, such as radioactivity. This corresponds to the passage from α to β as a 'giant.' A stage is reached in the region A, B when there is a balance and after that the radiation losses exceed the gains from contraction and other causes and the star cools, all the while its density increasing. This corresponds to the passage from β to γ as a 'dwarf.' The spectrum recedes from O to R; in time the star becomes too cold to radiate light and is then lost to us.

Eddington's theory of the interior of a star.

This theory can only be based upon the most general physical principles. These are, that the pressure of a gas of given density varies as its temperature (Boyle-Charles); the emission and absorption powers of a substance are equal (Kirchhoff); radiation pressure is determined by conservation of momentum (Maxwell); and the density of radiation varies as the fourth power of the temperature (Stefan). The gas constant (R) used in the Boyle-Charles equation depends upon the molecular weight assumed for the star. If the ionisation due to high temperature is so great that the n electrons revolving round the nucleus (see § 136) are detached, there will be $(n + 1)$ independent particles (the electrons and nucleus) so that if the term 'molecule' is used to denote ultimate particles (electrons or nuclei) the average molecular weight will be $\dfrac{A}{n+1}$ where A is the atomic weight. The atomic number (which is equal to the number n of the revolving electrons outside the nucleus of the neutral atom) is for most elements about one-half of the atomic weight A, so that $\dfrac{A}{n+1}$ is approximately 2. Eddington assumes this value 2 for the molecular weight inside the star, whatever its composition.

Using these principles and making an assumption concerning the mass-coefficient of absorption of radiation by the substance of the star, Eddington[1] shows that the effective temperature of a 'giant' varies as the sixth-root of the density, and that the total radiation is independent of the stage of evolution so long as the star is in a perfect gaseous condition, which is in agreement with the observed progress from α to β.

Eddington's theory has been further developed by E. A. Milne[2], who has recently applied the theory of Gibbs to the problem[3].

The above theory has since been modified and the whole question of stellar evolution is in an uncertain state.

The radiation theory outlined in this chapter is given in full detail in Planck's book ' *Wärmestrahlung* ' (chapters 2, 3 of the fifth edition)[4]; and the star theory (with an account of the radiation theory) in Eddington's book ' *The Internal Constitution of the Stars.* '[5][6]

---

[1] *Monthly Notices of the R.A.S.* p. 16, Nov. 1916; p. 596, June, 1917.
[2] *Phil. Mag.* 1924.　　　　　[3] *Camb. Phil. Soc. Proc.* 1925.
[4] Leipzig (J. A. Barth), 1923.　　[5] Cambridge (University Press), 1927.
[6] See also A. S. Eddington, '*Stars and Atoms,*' Oxford, 1927.

CHAPTER XV

THE THIRD LAW OF THERMODYNAMICS

164. *Kirchhoff's equation.*

This is an equation connecting the heat of reaction (in a process involving chemical or physical change) with the specific heats of the system before and after the change.

If Q is the heat evolved when the process occurs at temperature t without change of volume, c is the specific heat of the system before the change (reactants) and c' that of the system after the change (resultants), Kirchhoff's equation is

$$\frac{dQ}{dt} = c - c'.$$

To show this consider two states of the system

(1) in which the reactants are at temperature t,

(2) in which the resultants are at temperature $t + dt$.

The passage from state (1) to state (2) can be effected in two ways:

(*a*) the reactants can be heated from t to $t + dt$ and the reaction can then take place at temperature $t + dt$, so that the resultants are formed at that temperature with evolution of heat $Q + dQ$, or

(*b*) the reaction can take place at temperature t, so that the resultants are formed at that temperature with evolution of heat Q, and then the resultants can be heated from t to $t + dt$.

On account of the constancy of volume no external work is done, so that by the first law, the heat evolved is equal to the loss of internal energy (§ 10); and this loss is the same in processes (*a*) and (*b*) as each causes the passage from state (1) to state (2). Therefore the heat evolved in each process is the same.

Equating these, we have

$$Q + dQ - c\,dt = Q - c'\,dt,$$

so that
$$\frac{dQ}{dt} = c - c'.$$

If E is the *increase* of internal energy of the system due to the reaction, $E = -Q$, since no external work is done, so that

$$\frac{dE}{dt} = c' - c.$$

165. *Nernst's heat theorem*.

The Thomson equation

$$E = \psi - t\left(\frac{\partial\psi}{\partial t}\right)_v, \ (\S 55), \quad\ldots\ldots\ldots\ldots(1)$$

gives a relation between the internal energy E and the free energy ψ. Since the equation is linear E, ψ may represent differences of the internal and free energies of two given states, and will be so used in what follows. Nernst's object was to determine ψ when E is known from experiment in terms of t.

An arbitrary constant must arise when the Thomson equation is integrated to find ψ and the determination of this constant is the essence of the new theorem (*Nernst's heat theorem*). It adds to the subject a new axiom, which is the third law of thermodynamics.

Nernst assumed that the system is such that it can be cooled to the absolute zero continuously, that is, without any part of it undergoing a complete change at some temperature, such as freezing or transformation from one allotropic modification to another. This means that the specific heat of the system is a continuous function of the temperature t, and may be expressed by a power series in t.

Thus $\dfrac{dE}{dt} = c' - c$ means that $\dfrac{dE}{dt}$ is a power series in t, or

$$E = E_0 + \alpha t + \beta t^2 + \gamma t^3 + \ldots \ \ldots\ldots\ldots(2),$$

where $\alpha, \beta, \gamma, \ldots$ are known from experimental determinations of c, c'.

Therefore, from (1),

$$\psi - t\frac{d\psi}{dt} = E_0 + \alpha t + \beta t^2 + \gamma t^3 + \ldots,$$

$$\therefore \; -t^2\frac{d}{dt}\left(\frac{\psi}{t}\right) = E_0 + \alpha t + \beta t^2 + \gamma t^3 + \ldots,$$

$$\therefore \; \frac{d}{dt}\left(\frac{\psi}{t}\right) = -\frac{E_0}{t^2} - \frac{\alpha}{t} - \beta - \gamma t - \delta t^2 - \ldots,$$

$$\therefore \; \frac{\psi}{t} = A + \frac{E_0}{t} - \alpha \log t - \beta t - \frac{\gamma t^2}{2} - \frac{\delta t^3}{3} - \ldots$$

or $\quad \psi = E_0 + At - \alpha t \log t - \beta t^2 - \frac{\gamma t^3}{2} - \frac{\delta t^4}{3} - \ldots\ldots(3),$

where A is an unknown constant of integration.

From (2) and (3) we have

$$\left.\begin{aligned}
\frac{dE}{dt} &= \alpha + 2\beta t + 3\gamma t^2 + \ldots \\
\frac{d\psi}{dt} &= A - \alpha\,(1 + \log t) - 2\beta t - \frac{3}{2}\gamma t^2 - \frac{4}{3}\delta t^3 - \ldots
\end{aligned}\right\}(4).$$

From (2), (3), (4) it is seen that as $t \to 0$,

$$\psi \to E_0 \text{ and } \frac{d\psi}{dt} \to \infty,$$

while $\qquad\qquad E \to E_0 \text{ and } \frac{dE}{dt} \to \alpha.$

It would be natural to expect ψ, like E, to show a simple behaviour at low temperatures and not to produce this anomaly $\frac{d\psi}{dt} \to \infty$. This anomaly can be removed by assuming that $\alpha = 0$, for then the logarithm term in $\frac{d\psi}{dt}$ which produces the infinity is absent. Hence (4) becomes

$$\left.\begin{aligned}
\frac{dE}{dt} &= 2\beta t + 3\gamma t^2 + \ldots \\
\frac{d\psi}{dt} &= A - 2\beta t - \frac{3}{2}\gamma t^2 - \ldots
\end{aligned}\right\}\quad\ldots\ldots\ldots(5).$$

Nernst's idea is that E, ψ behave similarly near the absolute zero, so that $A = 0$ too.

Thus
$$\frac{dE}{dt} = 2\beta t + 3\gamma t^2 + \dots$$
$$\frac{d\psi}{dt} = -2\beta t - \frac{3}{2}\gamma t^2 - \dots \qquad\qquad \dots\dots\dots(6)$$

and if
$$E = E_0 + \beta t^2 + \gamma t^3 + \delta t^4 + \dots$$
then
$$\psi = E_0 - \beta t^2 - \frac{1}{2}\gamma t^3 - \frac{1}{3}\delta t^4 - \dots \qquad \dots\dots(7).$$

166. *The transformation of rhombic into monoclinic sulphur.*
$$\left.\begin{array}{l} E = E_0 + \beta t^2 \\ \psi = E_0 - \beta t^2 \end{array}\right\} \text{ approx.}$$

$\dfrac{dE}{dt} = c' - c = 2\beta t$, where c, c' are the specific heats before and after the change.

From measurements of c, c', it is found that $\beta = 1\cdot15 \times 10^{-5}$.

From observations of the heat of transformation, E_0 can be found, the formula $E = E_0 + \beta t^2$ being used.

The result for the range of t used was
$$E = 1\cdot57 + (1\cdot15 \times 10^{-5})\, t^2.$$
Therefore
$$\psi = 1\cdot57 - (1\cdot15 \times 10^{-5})\, t^2.$$

At the transition temperature from the one allotropic form to the other there is equilibrium between the two forms, which means that the usual ζ ($\equiv \psi + pv$) is the same for the two (§ 57). Here the change of pv is negligible, so that the usual ψ is the same for the two. In the formulae of this chapter, ψ means the *change* of the usual ψ due to the transformation; so that our present ψ is equal to zero when the transition occurs.

This leads to
$$t^2 = \frac{1\cdot57}{1\cdot15 \times 10^{-5}},$$
whence
$$t = 369\cdot5.$$

The observed temperature is $368\cdot4$ abs., which is in good agreement with the theory.

167. *Formulae for low temperatures.*

The formulae just used cannot hold down to low temperatures, as the Debye t^3-law (§ 147), confirmed by Eucken's experiments, requires the specific heat at low temperatures to vary as t^3, or $\dfrac{dE}{dt}$ varies as t^3.

This means that $\beta = 0$, $\gamma = 0$, from (6), so that (7) become

$$\left. \begin{array}{l} E = E_0 + \delta t^4 + \ldots \\[2mm] \psi = E_0 - \dfrac{1}{3}\,\delta t^4 - \ldots \end{array} \right\} .$$

A full account of the many and various applications of the theorem of which the one given above is an illustration has been given by Nernst in his book '*The New Heat Theorem.*'[1]

[1] English translation (Methuen), 1926.